RA 645.5 .A54 1999
Angle, James.
Occupational safety and
health in the emergency

LIBRARY
THE UNIVERSITY OF TEXAS
AT BROWNSVILLE
Brownsville, Tx -4991

OCCUPATIONAL SAFETY AND HEALTH IN THE EMERGENCY SERVICES

Online Services

Delmar Online
For the latest information on Delmar Publishers new series of Fire, Rescue and Emergency Response products, point your browser to:

http://www.firesci.com

Online Services

Delmar Online
To access a wide variety of Delmar products and services on the World Wide Web, point your browser to:

http://www.delmar.com
or email: info@delmar.com

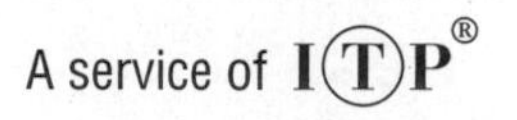

OCCUPATIONAL SAFETY AND HEALTH IN THE EMERGENCY SERVICES

James Angle

Delmar Publishers

an International Thomson Publishing company I(T)P®

Albany • Bonn • Boston • Cincinnati • Detroit • London • Madrid
Melbourne • Mexico City • New York • Pacific Grove • Paris • San Francisco
Singapore • Tokyo • Toronto • Washington

LIBRARY
THE UNIVERSITY OF TEXAS
AT BROWNSVILLE
Brownsville, Tx 78520-4991

NOTICE TO THE READER

Publisher does not warrant or guarantee any of the products described herein or perform any independent analysis in connection with any of the product information contained herein. Publisher does not assume, and expressly disclaims, any obligation to obtain and include information other than that provided to it by the manufacturer.

The reader is expressly warned to consider and adopt all safety precautions that might be indicated by the activities described herein and to avoid all potential hazards. By following the instructions contained herein, the reader willingly assumes all risks in connection with such instructions.

The publisher makes no representations or warranties of any kind, including but not limited to, the warranties of fitness for particular purpose or merchantability, nor are any such representations implied with respect to the material set forth herein, and the publisher takes no responsibility with respect to such material. The publisher shall not be liable for any special, consequential, or exemplary damages resulting, in whole or in part, from the readers' use of, or reliance upon, this material.

Cover photo courtesy of Bill Tompkins, NJ Metro Fire Photographers

Delmar Staff
Publisher: Alar Elken
Acquisitions Editor: Mark Huth
Developmental Editor: Jeanne Mesick
Project Editor: Megeen Mulholland
Production Coordinator: Toni Bolognino
Art and Design Coordinator: Michele Canfield
Editorial Assistant: Dawn Daugherty

COPYRIGHT © 1999
By Delmar Publishers

an International Thomson Publishing company I(T)P®

The ITP logo is a trademark under license

Printed in the United States of America

For more information, contact:

Delmar Publishers
3 Columbia Circle, Box 15015
Albany, New York 12212-5015

International Thomson Publishing Europe
Berkshire House
168-173 High Holborn
London, WC1V7AA
United Kingdom

Nelson ITP, Australia
102 Dodds Street
South Melbourne,
Victoria, 3205 Australia

Nelson Canada
1120 Birchmont Road
Scarborough, Ontario
M1K 5G4, Canada

International Thomson Publishing France
Tour Maine-Montparnasse
33 Avenue du Maine
75755 Paris Cedex 15, France

International Thomson Editores
Seneca 53
Colonia Polanco
11560 Mexico D. F. Mexico

International Thomson Publishing GmbH
Königswinterer Straße 418
53227 Bonn
Germany

International Thomson Publishing Asia
60 Albert Street
#15-01 Albert Complex
Singapore 189969

International Thomson Publishing Japan
Hirakawa-cho Kyowa Building, 3F
2-2-1 Hirakawa-cho, Chiyoda-ku,
Tokyo 102, Japan

ITE Spain/ Paraninfo
Calle Magallanes, 25
28015-Madrid, Espana

All rights reserved. No part of this work covered by the copyright hereon may be reproduced or used in any form or by any means—graphic, electronic, or mechanical, including photocopying, recording, taping, or information storage and retrieval systems—without written permission of the publisher.

1 2 3 4 5 6 7 8 9 10 XXX 04 03 02 01 00 99

Library of Congress Cataloging-in-Publication Data

Angle, James.
Occupational safety and health in the emergency services / James Angle.
p. cm.
Includes bibliographical references and index.
ISBN 0-8273-8359-2 (hc.)
1. Emergency medical personnel—Health and hygiene. 2. Fire fighters—Health and hygiene. I. Title.
RA645.5.A54 1998
362.18—dc21 98-28536
CIP

This book is dedicated to all those emergency responders who have been injured or killed in the line of duty and the valuable lessons they have taught us

Contents

Foreword

The book you are about to read is written about a significant change in our emergency response business that involves a major redefinition of how we manage the welfare and survival affairs of our human assets. We are now asking (actually requiring) our traditional systems, and the managers that go with those systems, to implement and refine an occupational health and safety program that effectively protects our workers.

This is a huge shift from the way we have done business in the past. In the old days we managed the troops a lot like we handled our apparatus and equipment. We purchased it, we put it in service, we used it, and when we were finished with it, we surplused it. In those days, the biggest difference between how we managed the humans and the "stuff" was that we actually spent a lot more time, effort, and resources maintaining the hardware than we did the humans. In fact, we were a lot more attracted to supporting the equipment because it did whatever we directed it to, it never talked back, and it always behaved in a way we could understand. For the most part, we used an old time vertical military model to manage our personnel. The model was basically designed to conduct the organizational activities of waging war. The model, and the mentality that went with it, assumed that if you were in the war business, some folks would get scuffed up (or worse). The system backend-loaded a benevolent set of responses that supported the basic fatalistic assumption of worker injury and death—things like big insurance plans, worker's compensation benefits, disability pensions, and huge funeral festivals. Those are (and probably always will be) very logical and legitimate benefits for workers who go into highly hazardous areas to do their jobs. The problem in the past was that these benefits were the major focus of how we responded to protecting the workers. These benefits in a well-managed system will actually compliment the occupational health and safety approach we use today.

The old backend-loading approach was directed toward reacting after something bad had happened. The new approach predicts and attempts to prevent the ugly stuff from ever occurring. Like any other organizational change, the new occupational injury/death prevention routine requires an old system to do new things, and it will take a while for us to become comfortable and skillful with the new program. This book provides a broad overview of the different parts of a complete health and safety program.

Implementing those program components provides a game plan for the reactive to proactive shift. This change is particularly difficult for us, simply because our historic approach is to react quickly to the customer's problem. Nothing in this book suggests that we change our commitment to the fast and effective delivery of service. That commitment is why we are and always will be in business. What this new routine does is create a sensible balance that also considers the responder as a customer.

This new balance recognizes that it is impossible for us to effectively deliver service if we are not protected from whatever is causing the customer a bad day. The health and safety program requires that we take a long-term, before/during/after, view of worker welfare.

In my travels, I hear many leaders say that their people are the most important asset in their organizations. Now when I hear someone say that, I think to myself, "OK, prove it." A workforce that routinely must put its bodies directly between the customers and their hazardous problems, has a very special set of welfare needs. How well we implement the program outlined in this book will determine in the most basic and important way if we can indeed "prove it."

Chief Alan V. Brunacini
Phoenix Fire Department

Preface

Each year, a number of emergency responders are killed or injured in the line of duty. It is my hope that this book will have a significant, positive impact on these statistics. Emphasis on occupational safety and health in fire and emergency medical services (EMS) is a fairly new development. Increasing emphasis and an increase in accountability for safe practices over the last 10 years have afforded the opportunity for a book that focuses on both fire and EMS safety and health issues from a program management approach. Since the safety trend has emerged, many colleges have begun to offer accredited courses in this area as part of degree programs or as stand-alone courses.

ABOUT THIS BOOK

This book focuses on a comprehensive approach to emergency service occupational safety and health. The purpose of a book of this type is to provide instructors, students, safety officers, fire and EMS department managers, and others in the emergency service field a one-stop resource for safety program management.

The text introduces readers to occupational safety and health from a historical standpoint and guides them through a process that will allow them to put the information to work within their organizations. The outcome will be a safer work environment, which will reduce costs and increase productivity

HOW TO USE THIS BOOK

This text is designed to be used for a semester or quarter-hour college level course on occupational safety and health within the fire or EMS discipline. The book was also designed to be used by fire and EMS agencies, whether fully paid, combination, or volunteer, as an in-house reference, a manual for an organizational training program, and may be used as a promotional testing reference. Although the text does not go into great detail on every safety issue, it is designed to give the reader a broad overview of issues to consider for various types of incidents common to emergency response. Each chapter begins with a case study designed to create thought and stimulate discussion. The chapter then takes the reader through the subject in a logical manner. Finally, each chapter ends with review questions designed to allow the reader to assess the knowledge they have learned based on the objectives of the chapter. Also included at the chapter end are activities designed to allow the reader to put the concepts discussed in the chapter to work within their own organizations.

Chapter 1 provides a brief history into occupational safety and health for emergency services and brings the reader to an understanding of the importance and the rationale for the increasing emphasis over the last 10 years. Chapter 2 defines and reviews standards and regulations that have an impact on safety programs. Risk analysis and management are presented in Chapter 3 to allow the reader to form an understanding of this important concept before specific situations are presented. Specific safety issues are described in Chapters 4 through 8: Preincident safety; fire, medical/rescue, and specialized incidents; and postincident implications are discussed. Chapter 9 focuses on issues relating to personnel roles and responsibilities. Personnel issues are often the most difficult to deal with in terms of safety programs. The remaining four chapters deal with program management including development and ongoing management, evaluation of effectiveness, information management, and current issues and future trends.

Key terms and notes are included throughout the chapters to highlight important points and to introduce new concepts. Each chapter lists a variety of learning objectives; these should be reviewed prior to reading the chapter.

ABOUT THE AUTHOR

James S. Angle is the fire chief of the Palm Harbor Fire-Rescue Department in Pinellas County, Florida. The department protects 62,000 people in a 20-square-mile area operating from four fire stations, providing a full range of services, including fire prevention, public education, advanced life support, rescue, hazmat, and fire suppression. He is a 23-year emergency service veteran, having begun his career in the Monroeville Fire Department in suburban Pittsburgh, Pennsylvania, and the Pittsburgh EMS Bureau. His background includes employment with five emergency service agencies both small and large, working through the ranks from firefighter/paramedic to his current position as fire chief.

His education includes an associate degree in Fire Science Administration from Broward Community College, a bachelor's degree in Fire Science and Safety Engineering from the University of Cincinnati, and a master's degree in Business from Nova University. He also holds instructor certification in three areas from the Florida Bureau of Fire Standards and Training, is certified as a paramedic, and is a graduate of the Executive Fire Officer Program of the National Fire Academy.

As an instructor, he teaches in the Public Services, fire and EMS, division of Edison Community College, as well as Hillsborough Community College. He has taught for the University of Cincinnati in the Fire Science program and has delivered seminars to national audiences at three Fire Department Instructors' Conferences. He has also published articles in two fire service journals.

ACKNOWLEDGMENTS

This section could be a chapter in itself. I would be remiss if I did not begin with thanks to my wife, Joann, and my two sons, Anthony and Austin. Without their support and willingness to give up Daddy for many weekends and evenings, this book would not have been possible.

I would like to thank my father, who passed away during the writing of this book, as he is the one who took me to my first fire, which made me realize that I never wanted to do anything else other than be a firefighter.

I could name many individuals who impacted me throughout my career, but the list would be longer than the book. So many have helped me to make this my career, and not just a job. I would like to recognize the Board of Fire Commissioners and the members of the Palm Harbor Fire Rescue Department who have provided encouragement and support throughout this endeavor. I also thank the management and members of the South Trail Fire Department, where for 8 years I served as training and safety officer and where I realized that through a good process, management support, and member support, an organization can reduce injuries.

A special thanks to Mark Huth from Delmar who took a chance and allowed me to make another dream a reality. To Jeanne Mesick of Delmar who provided support, understanding, and friendship throughout this project.

Acknowledgments

The author and Delmar Publishers would like to acknowledge and thank the following people who offered valuable advice that improved the quality of this book:

Chief Alan V. Brunacini
Phoenix Fire Department

Lee Cooper
Wisconsin Indianhead Technical College
New Richmond, WI

Rudy Horist
Elgin Fire Department
Elgin, IL

Robert Klinoff
Kern County Fire Department
Bakersfield, CA

Robert Laeng
Training for Life Safety
Bulls Gap, TN

Randy Lawton
Elmwood Township Fire Rescue
Traverse City, MI

Andy Mancusi
Hawthorne Fire Department
Hawthorne, NY

Jack Reed
Indiana University of Pennsylvania
Indiana, PA

Bill Shouldis
Philadelphia Fire Department
Philadelphia, PA

Introduction to Emergency Services Occupational Safety and Health

Learning Objectives

Upon completion of this chapter, you should be able to:

- Discuss the history of emergency service safety and health programs.
- Identify, by using historical data, the safety and health problem as it is today.
- Describe the efforts that have been made to address the safety and health problem among emergency service occupations.
- List the national agencies that produce annual injury and fatality reports for emergency services.
- Identify the information that can be obtained from annual injury and fatality reports.

CASE REVIEW

The new chief arrived at the midsized fire department where he was assuming command. The department seemed to be running very well, and the members were eager for some new leadership. Having a reasonably good background in firefighter/emergency medical services (EMS)-related safety and health concerns, the chief was quick to note high workers' compensation premiums and a high injury rate.

One early task for the chief was to identify why this rate was high and to find ways to lower the injury rate, the lost time, and the workers' compensation premiums. The chief quickly sought the assistance of the training/safety officer. Their first step was to analyze injuries for the past two years. This data was then compared to the most recent National Fire Protection Association (NFPA) report on injuries and the most recent report from the International Association of Fire Fighters (IAFF) report on injuries. Although the department fared well in most of the categories, it become clear that the rate of back injuries and sprain and strain was extremely high. Further, the incidence of injury at the fireground was almost zero, while in other on-duty areas it was very high.

The department initiated a strong program to try to reduce these types of injuries. Included in the program were more aggressive injury prevention activities from the safety committee including posters, reminders, and a training program on proper lifting. Furthermore, the department developed a return-to-work process designed to make sure employees, after injury, are ready to return to duty.

During the study period in 1996, there were forty total injuries reported to the workers' compensation carrier. During the same period in 1997, there were seventeen total reported injuries, a greater than 50% improvement. However, the work of the chief and the department is not over. For the size of the department, even seventeen is a number that can be reduced. Using some of the techniques described in this textbook and national statistics, the department continues to improve.

INTRODUCTION

Firefighting is sometimes described as the nation's most dangerous occupation because of the high rate of acute and chronic injuries and deaths. Firefighters are exposed to chemicals, carcinogens, extreme weather, building collapses, smoke, heat, and traveling under less-than-ideal roadway conditions. It is no wonder that the occupation suffers a high rate of injury and death.

Emergency medical service responders face many of the same hazards as firefighters. In fact, emergency medical responders face additional exposure, including concerns such as infectious diseases. Most firefighters are also responsible for emergency medical response. In fact, the fire service is the largest provider of pre-hospital care (Ludwig 1998).[1] Because of this close relationship between the firefighters and the responders to emergency medical incidents, a complete occupational safety and health program has to focus on both fires and

[1] Ludwig, Gary. *On Scene*, May 8, 1998. International Association of Fire Chiefs.

Figure 1-1 *Early fire apparatus design did not consider the safety of the responders.*

medical emergencies. Therefore, an examination of the problems and the methods to improve an historically poor record is conducted from the emergency service standpoint.

HISTORY OF EMERGENCY SERVICES SAFETY AND HEALTH

For many years, the poor injury and death rate for firefighters was accepted as part of the occupation (Figure 1-1). As firefighters moved into the role of EMS responders, the injury potential increased, not only because of additional exposures, but also because of increased incidents that lead to a higher potential for responding and returning accidents.

Many of the early texts used in the education and training of firefighters and EMS responders make little or no reference to the injury and death problem from a standpoint of improvement. There are references to the dangers of the profession; however, little is provided for prevention strategies.

One of the first publicized documents making reference to firefighter safety was *America Burning.*[2] In this 1973 report to the president of the United States, the National Commission on Fire Prevention and Control reported on the nation's fire problem (see Figure 1-2). One section of the report focused on firefighter safety from the perspective of staffing, education, and equipment. The 1980s saw a significant increased interest in the safety problem facing the emergency responders, and this trend has continued into the 1990s. One of the most controversial,

[2] National Commission on Fire Prevention and Control, 1973. Washington, D.C.

The Commission on Fire Prevention and Control has made a good beginning, but it cannot do our work for us. Only people can prevent fires. We must become constantly alert to the threat of fires to ourselves, our children, and our homes. Fire is almost always the result of human carelessness. Each one of us must become aware—not for a single time, but for all the year—of what he or she can do to prevent fires.

—President Richard M. Nixon
September 7, 1972

Figure 1-2 *President Nixon's preface to "America Burning."*

NFPA 1500
the National Fire Protection Association's consensus standard on fire department occupational safety and health

yet positive steps, in the safety and health area was the publication of **NFPA 1500,** the Fire Department Occupational Safety and Health Program published in 1987. This standard and the series of other safety standards that followed set the safety and health movement in motion. The standard was controversial because of potential changes in operations and costs, but it caused managers, unions, volunteer fire chiefs, and others to note that the statistics for death and injury were unacceptable, and something had to be done.

Emergency services organizations are often reluctant to make and accept change. However, procedures, techniques, and often equipment—do change. Often these changes are engineered without considering the human element of the change. For example, a department can implement the best all-hazard incident management system with scene accountability, but if individual companies or crews freelance within the system, the program will fail. All too often, the attitude of the response personnel mimics that of our nation's population when it comes to fires and injuries—*it won't happen to me*. However, it *does* happen to someone almost 100,000 times a year for injuries and almost 100 times a year for deaths.

■ Note
Another significant improvement since the movement began is the recognition that the local injury problem should be evaluated.

Since the publication of NFPA 1500, emergency services organizations have seen changes. It is now common to find organization charts that include a health and safety officer, new textbooks and new editions with sections on health and safety, safety and health committees formed with representation from all levels of the organization, and standard operational procedures intended to provide a safe working environment. Incident management systems have been established, the use of personal protective equipment is required, and safer equipment has been developed. All of this has been accomplished with the thought of improved safety and health for the responder. Many emergency service organizations offer yearly medical examinations in addition to the examination at time of hire. These regular physicals allow the formation of a baseline for an annual comparison of medical history over the length of an employee's career. This trend will and should continue. Another significant improvement since the movement began is the recognition that the local injury problem should be evaluated. National statistics have been gathered on occupational injuries, but often this information is not compared to local statistics until a problem or tragic event occurs.

IDENTIFICATION OF THE SAFETY PROBLEM

A number of resources are available to help a safety and health officer develop a program. Clearly, local program design and development must focus on local problems. It would not make sense for Florida departments to invest considerable money to prevent injuries associated with exposure to cold. Likewise, a department providing EMS transport that experiences a high number of back injuries may want to make a substantial investment in a back injury prevention program. Local statistics can be gathered, and the data can be used to determine the local safety and health problem. Techniques for gathering this information and the process for the analysis are provided in Chapter 12. However, local data should be compared to that of the larger population from the state and national figures. The following organizations publish annual safety and health data that can be useful in this endeavor.

National Fire Protection Association

In addition to publishing standards, the NFPA publishes annual reports on both occupational injuries and deaths in the fire service in its *Fire Journal* magazine. The NFPA has been compiling these reports since 1974. The NFPA death survey is a report on all firefighter deaths and includes analysis of the fatalities in terms of type of duty, cause of death, age group comparisons, and population-served comparisons. The NFPA injury survey is not an actual survey of all departments but is a sample used to project the national firefighter injury experience. Although a prediction, the survey inspires a high level of confidence and is representative of all sizes and types of departments. These reports are very comprehensive and useful for the safety program manager to compare national to local experience.

United States Fire Administration
department under the Federal Emergency Management Agency that directs and produces fire programs, research, and education

National Fire Incident Reporting System (NFIRS)
uniform fire incident reporting system for the United States; the data from this report is analyzed by the United States Fire Administration

United States Fire Administration

The **United States Fire Administration** (USFA) oversees the **National Fire Incident Reporting System** (NFIRS). Part of the NFIRS report is for collecting data for firefighter casualties. This data can be obtained through the USFA and can be used to compare national data to local data. However, there are some problems with this system. The NFIRS is a voluntary system in which departments may or may not participate, which creates the problem of not knowing what injuries or deaths may go unreported because a local jurisdiction did not participate in the program. Further, as the name implies, the report is used for incidents; therefore, injuries that occur to on-duty personnel outside of incidents is not captured.

The NFIRS data is formulated into reports in various formats and is available by mail from the USFA publication center. Much emergency service safety and health data, including case analyses of incidents where fatalities or injuries occurred and protective clothing field test results, are also available from the USFA publication center. The USFA also has an intensive worldwide web page providing a means to order publications, but many are downloadable and can be reviewed immediately. The 1996 firefighter fatality report from the USFA is included as Appendix I. The Internet address for the USFA can be found in Appendix II, or the USFA can be contacted by telephone at 1-800-238-3358.

International Association of Fire Fighters

International Association of Fire Fighters
labor organization that represents the majority of organized firefighters in the United States and Canada

Since 1960, the **International Association of Fire Fighters** (IAFF) has produced an annual firefighter injury and death study (see Figure 1-3). This research reports on a number of issues relating to safety and health similar to that of the NFPA.

Figure 1-3 *The International Association of Fire Fighters publishes an annual firefighter injury and death study.*

However, the report is more in-depth and provides information on lost-time injuries and exposure to infectious diseases. This report is another resource for the health and safety program manager, but the data is only gathered from paid fire departments that have IAFF affiliation, which limits the study to these departments. The IAFF, along with the **International Association of Fire Chiefs** (IAFC), in 1997 presented the Fire Service Joint Labor/Management Wellness-Fitness initiative. This initiative contains a component of data reporting to the IAFF on employee wellness, fitness, and injury issues. This report will be a great resource to the safety manager when sufficient data is received, but it will only contain data from departments that choose to participate.

International Association of Fire Chiefs (IAFC) organization of fire chiefs from the United States and Canada

Occupational Safety and Health Administration

The **Occupational Safety and Health Administration** (OSHA) is an agency within the Department of Labor. OSHA regulations are applicable to many public fire departments and to all private fire departments (see Chapter 2 for additional information on OSHA regulations applicable to public agencies in your state). OSHA requires certain reporting requirements for occupation-related injuries and deaths. This requirement for record keeping and reporting allows OSHA to compile useful statistics and to study causes of occupational injuries in order to develop prevention strategies and countermeasures. Because not every state requires that public fire departments comply with OSHA regulations, this data also is incomplete.

Occupational Safety and Health Administration federal agency tasked with the responsibility for the occupational safety of employees

REVIEW OF CURRENT NATIONAL INJURY STATISTICS

To provide readers of this text with benchmarks and an insight into the type of information that can be obtained from an annual injury report, we examine the 1996 NFPA Firefighter Injury report. The NFPA report is usually published late in the year following the year studied. For example, the 1996 injury report was published in the November/December 1997 issue of *Fire Journal*. As described previously, this report is a statistical prediction of the national injury experience. The report examines the injury experience from a number of perspectives, listed below. Each report is useful in analyzing and comparing the local problem with the national problem.

Injuries by Type of Duty The classification is divided into five subcategories (see Figure 1-4):

- Responding/returning
- Fireground
- Nonfire emergencies
- Training
- Other on-duty

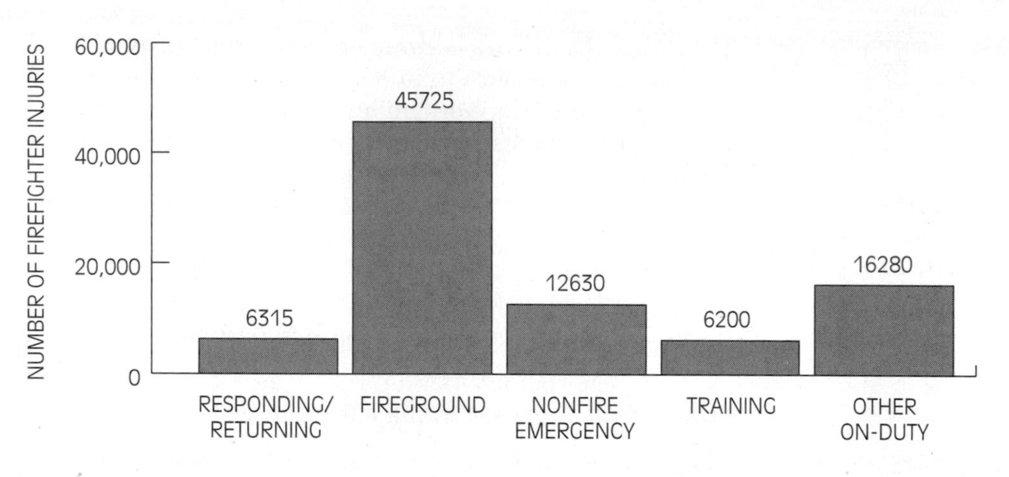

Figure 1-4 *Firefighter injuries by type of duty. (Reprinted with permission from* NFPA Journal, *November/ December 1997, National Fire Protection Association, Quincy, MA 02269.)*

In 1996, as in all other years, fireground injuries accounted for the highest number of injuries.

Nature of Injuries Nature of injury (see Figure 1-5) is subdivided into ten classifications, which are reported by type of duty as well.

- Burns (fire or chemical)
- Smoke or gas inhalation
- Other respiratory distress
- Eye irritation
- Wound, cut, bleeding, bruise
- Dislocation, fracture
- Heart attack or stroke
- Strain, sprain, muscular pain
- Thermal stress (frostbite, heat exhaustion)
- Other

Fireground Injuries by Cause As fireground injuries historically cause the highest number of injuries per type of duty, fireground injuries by cause are also reported (see Figure 1-6). The nine categories for fireground injuries by cause are:

- Struck by object
- Stepped on, contact with object
- Extreme weather

Firefighter Injuries by Nature of Injury and Type of Duty, 1996

	Responding to or Returning from an Incident		Fireground		Nonfire Emergency		Training		Other On-Duty		Total	
Nature of Injury	**Number**	**Percent**	**Number**	**Percent**	**Number**	**Percent**	**Number**	**Percent**	**Number**	**Percent**	**Number**	**Percent**
Burns (Fire or Chemical)	65	1.0	4,360	9.5	140	1.1	635	10.2	215	1.3	5,415	6.2
Smoke or Gas Inhalation	115	1.8	4,660	10.2	305	2.4	70	1.1	105	0.6	5,255	6.0
Other Respiratory Distress	45	0.7	740	1.6	210	1.6	75	1.2	125	0.8	1,195	1.4
Eye Irritation	225	3.6	2,735	6.0	390	3.1	165	2.7	620	3.8	4,135	4.7
Wound, Cut, Bleeding, Bruise	1,375	21.8	8,775	19.2	2,055	16.3	1,085	17.5	3,325	20.4	16,615	19.1
Dislocation, Fracture	190	3.0	1,090	2.4	260	2.0	235	3.8	550	3.4	2,325	2.7
Heart Attack or Stroke	25	0.4	300	0.7	45	0.4	35	0.6	310	1.9	715	0.8
Strain, Sprain, Muscular Pain	3,545	56.1	17,455	38.2	7,020	55.6	3,160	51.0	8,540	52.5	39,720	45.6
Thermal Stress (frostbite, heat exhaustion)	60	1.0	2,720	5.9	185	1.5	260	4.2	100	0.6	3,325	3.8
Other	670	10.6	2,890	6.3	2,020	16.0	480	7.7	2,390	14.7	8,450	9.7
	6,315		45,725		12,630		6,200		16,280		87,150	

SOURCE: NFPA's Survey of Fire Departments of U.S. Fire Experience (1996).
NOTE: If a firefighter sustained multiple injuries for the same incident, only the nature of the single most serious injury was tabulated.

Figure 1-5 *Nature of injuries and type of duty. (Reprinted with permission from* NFPA Journal, *November/December 1997, National Fire Protection Association, Quincy, MA 02269.)*

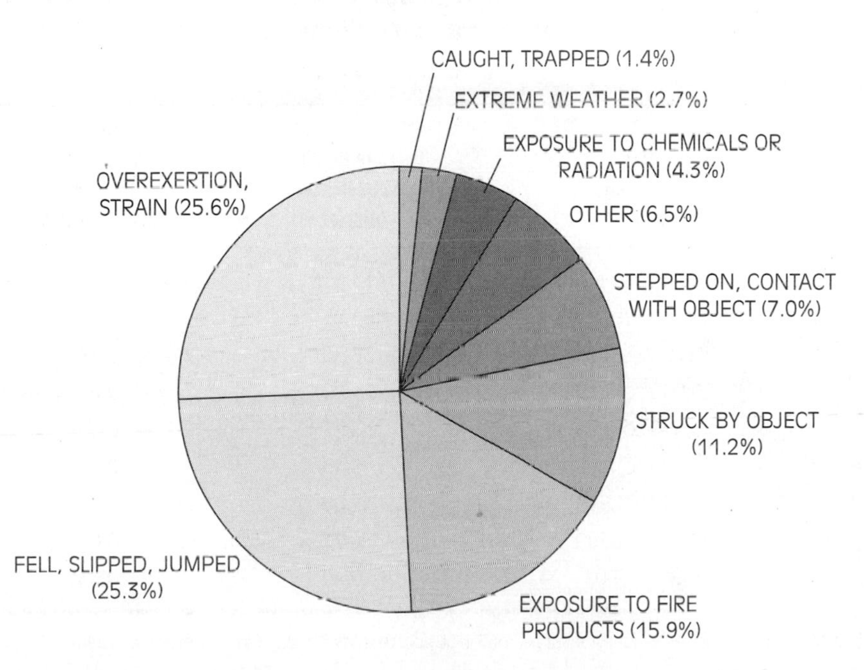

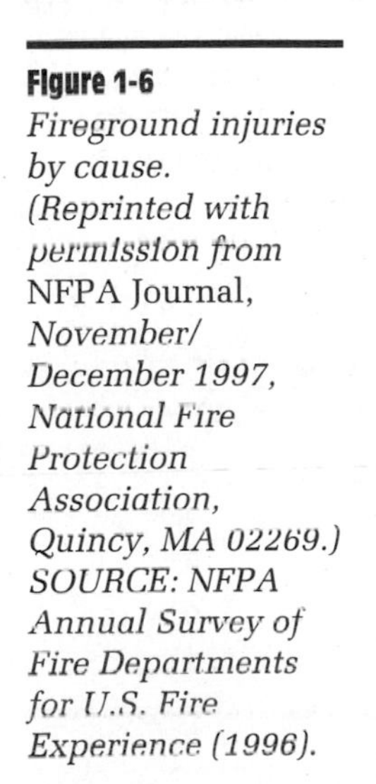

Figure 1-6 *Fireground injuries by cause. (Reprinted with permission from* NFPA Journal, *November/December 1997, National Fire Protection Association, Quincy, MA 02269.) SOURCE: NFPA Annual Survey of Fire Departments for U.S. Fire Experience (1996).*

- Caught, trapped
- Exposure to fire products
- Exposure to chemicals or radiation
- Fell, slipped, jumped
- Overexertion
- Other

■ Note
Examination of Figure 1-8 shows that while injuries have decreased in total, the rate of injuries per 1,000 fires has remained relatively constant.

Average Injuries by Size of Population Protected Several items relate to this category and are useful in analysis (Figure 1-7). Not only are the average number of injuries per population reported, but they are further divided to show the number that required hospitalization, and the number that required hospitalization per 100 injuries.

■ Note
Review of this annual report must be required for safety and health program managers.

Also of interest in the report is the ten-year report on total firefighter injuries per year. Examination of Figure 1-8 shows that while injuries have decreased in total, the rate of injuries per 1,000 fires has remained relatively constant. The NFPA report further summarizes the data presented in Figures 1-4 through 1-8 in the text of the report. The report also provides an overview and lengthy reports on selected incidents. Review of this annual report must be required for safety and health program managers.

Average Number of Firefighter Injuries and Injuries Requiring Hospitalization per Department, by Population of Community Protected for All Types of Duty, 1996

Population of Community Protected	Average Number of Firefighter Injuries	Average Number of Injuries Requiring Hospitalization	Number of Injuries Requiring Hospitalization per 100 Injuries
500,000 to 999,999	433.10	9.22	2.13
250,000 to 499,999	151.25	2.78	1.84
100,000 to 249,999	56.30	2.53	4.49
50,000 to 99,999	21.00	0.68	3.24
25,000 to 49,999	8.77	0.19	2.17
10,000 to 24,999	3.15	0.17	5.40
5,000 to 9,999	1.48	0.12	8.11
2,500 to 4,999	0.77	0.07	9.09
Under 2,500	0.40	0.03	7.50

SOURCE: NFPA's Survey of Fire Departments for U.S. Fire Experience (1996).

Figure 1-7 *Injuries by population. (Reprinted with permission from* NFPA Journal, *November/December 1997, National Fire Protection Association, Quincy, MA 02269.)*

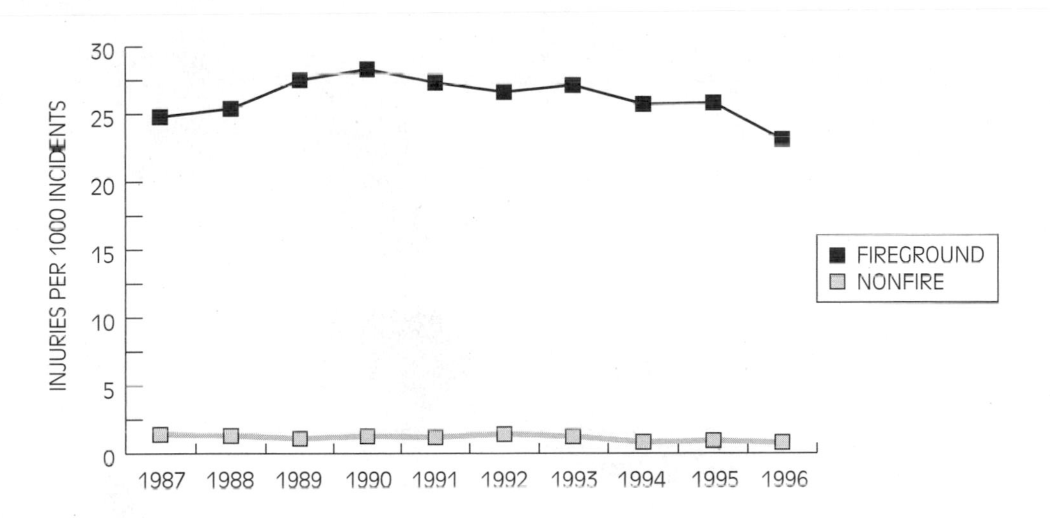

Figure 1-8 *Ten-year trends showing number of injuries per 1,000 fires. (Reprinted with permission from* NFPA Journal, *November/December 1997, National Fire Protection Association, Quincy, MA 02269.)*

WHAT IS BEING DONE

risk management processes and programs put in place to minimize risks and reduce the consequences when an accident occurs

As described previously, a number of changes have occurred since the publication of NFPA 1500. Departments are taking a more proactive approach to health and safety. **Risk management** is now a concept as common in the fire station as it is in private industry. What has been done over the last decade can probably fit nicely into the acronym SAFEOPS, which means:

Supervision. We have hired risk management specialists, assigned safety officers, empowered supervisors to be more safety aware, hired or contracted with fitness and wellness experts, and in many cases contracted with physicians to handle physicals and medical surveillance of department employees.

Attitude. Employees are made aware of the safety and health problem and empowered to be proactive concerning safety. When training begins, there is an emphasis placed on safety; in national journals, there are articles relating to safety and health almost monthly. By virtue of the resources put forth and the supervision focusing on safety and health, employees' attitudes toward safety are slowly improving. Attitude is the key to the success of any safety program; we must constantly seek improvement in attitude.

Fitness/wellness. Many departments have increased their commitment to fitness and wellness. Fire stations are equipped with workout equipment (see Figure 1-9) and in some cases departments are contracting with local gyms. Mental fitness has become a priority, with departments having critical incident stress debriefing teams and employee assistance programs. Some departments offer classes on weight reduction, smoking cessation, and eating right. Fitness and wellness are critical issues in the overall safety and health program.

Education. The level of safety education, both career and volunteer, for both entry-level and supervisory positions has increased, providing for a broad professional knowledge that improves the ability to collect, analyze, and present

Figure 1-9 *Fitness equipment should be provided in the station.*

data related to problems. Programs have been developed to better prepare employees to safely perform their jobs, such as how to lift correctly to prevent sprains and strains.

Organizational involvement. Employee-based safety and health committees, fitness committees, and labor/management relations committees support the safety and health programs by developing new procedures or analyzing new equipment. Employee suggestion programs for safety-related improvements allow all members of an organization to provide input and be stakeholders in making their jobs safer.

Procedures. Operational procedures have been adopted or changed to reflect the needs of safety and health programs. Included are incident management systems, personnel accountability procedures, use of personal protective equipment, and responding to nonemergency incidents in a nonemergency mode.

Standards/regulations. Standards and regulations have also changed to meet the needs of the occupation. The *Standard on Live Fire Training Evolutions* (**NFPA 1403**) was written in hopes of preventing future deaths and injuries related to live fire training. Federal agencies also have promulgated regulations aimed at reducing occupational injuries and deaths. The OSHA **blood-borne pathogen** regulation requires procedures and precautions to prevent the spread of blood-borne diseases. Many departments may have adopted these procedures even if they had not been forced to comply.

This is just an overview of initiatives and programs that have been developed to improve the safety and health of emergency responders. Many are not universally accepted or practiced. Hopefully, this text will provide readers with further insight into some of these programs and procedures and motivate them to seek implementation in their own departments.

NFPA 1403
National Fire Protection Association's consensus standard on live fire training evolutions

bloodborne pathogen
disease carried in blood or blood products

Summary

Response to emergencies is by nature a hazardous occupation. Historically, prior to the publication of *America Burning,* little attention was given to overall safety and health in this profession. However, in the 1980s, emphasis on emergency service provider safety increased significantly and responders saw the publication of NFPA 1500. It is unlikely, given the nature of the occupation, that all injuries and deaths could be totally eliminated. However, with good risk management, the frequency and severity could be reduced. The first step in designing a safety and health program is to identify and understand the problem, which can be accomplished through analysis of local data and comparisons with the national experience. Several organizations provide sources for data relating to safety and health programs. The National Fire Protection Association publishes an annual report of firefighter injuries and firefighter deaths, the United States Fire Administration compiles injury and fatality data through the use of the National Fire Incident Reporting System, the International Association of Fire Fighters produces an annual injury and death report, and OSHA collects and analyzes data relating to occupational injuries. Many changes have occurred in the last ten to fifteen years regarding emergency responder safety, health, and wellness. The changes include increased awareness and priority of safety issues for frontline supervisors, improvement in the safety attitude at all levels of the organization, increased emphasis on fitness and wellness, higher levels of education for entry-level employees and supervisors, involvement through committee and suggestion programs at all levels of the organization, adoption of revised procedures related to safety operations, and promulgation of standards and regulations designed to improve the safety and health of responders.

Concluding Thought: Improvements have been made to improve responders' safety and health. However, as presented in this chapter, the number of fires is down, but the injuries have remained constant. Therefore, we need better programs, research, data, and organizational commitment to improve even more. Each organization can and should benchmark itself against similar organizations to further improve the statistics.

Review Questions

1. What do the letters of the acronym SAFEOPS stand for?
2. Which of the following organizations survey every fire department, both paid and volunteer, in every state, for injury and death statistics?
   A. USFA
   B. IAFF
   C. NFPA
   D. OSHA
   E. None of the above

3. According to the chapter text, occupational safety and health moved ahead rapidly during which of the following decades?
   A. 1950s
   B. 1960s
   C. 1970s
   D. 1980s
4. Which major initiative did the IAFF and the International Association of Fire Chiefs work on together?
   A. NFPA 1500
   B. OSHA 1910.120
   C. The SAFEOPS approach
   D. a fitness/wellness program

Use Figures 1-4 through 1-8 to answer questions 5 through 10.

5. How many injuries per 1,000 fires occurred in 1996?
6. What percentage of injuries occurred on the fireground?
7. The most sprains and strains occurred during what type of duty?
8. What population range has the highest rate of injuries?
9. What population range has the highest rate of injuries that required hospitalization?
10. What caused the greatest percentage of fireground injuries?

Activities

1. Obtain a copy of the most recent death or injury report from any of the organizations listed above. Review the statistics and compare them to those of your department.
2. Using the SAFEOPS approach, analyze what your department has done since the publication of the 1987 edition of NFPA 1500 in terms of improving responder safety and health. List these changes and evaluate their effectiveness.

Review of Safety-Related Regulations and Standards

Learning Objectives

Upon completion of this chapter, you should be able to:

- Discuss the difference between regulations and standards.
- Discuss the concept of standard of care.
- List and discuss federal regulations that have an impact on safety and health programs.
- List and discuss the major National Fire Protection Association standards that have an impact on occupational safety and health programs.
- Discuss the Alliance for Fire and Emergency Management's performance standard for incident safety officer.
- Discuss the role of related regulations and standards and their safety and health implications.

CASE REVIEW

On December 6, 1992, a tragic accident occurred during a training exercise being held in Parsippany, New Jersey. This accident sent shock waves through the fire-rescue service, because the failures that occurred were the same as had occurred before in other training accidents.

A Firefighter I training program was being presented by the Greystone Park Fire Department during the fall of 1992. As part of this program, the participants were to be given the opportunity to experience a live fire. On the day of the accident the students, along with one instructor, arrived at the location where the fire was to be conducted. The live fire was to be set not in an approved burn building, but in a converted school bus. The bus had metal plates affixed to the windows leaving the only means of entry, exit, or ventilation through the front passenger door or the emergency exit. A couch was placed in the bus to be used as fuel for the fire. Three 55-gallon drums cut in half that had been used to produce smoke in previous drills were also in the bus at the time of the incident. According to a witness's account, a railroad flare was used to start the couch on fire at about 9:50 A.M. After allowing a preburn of about 5–10 minutes, students were taken into the bus. At this time no hose lines were taken into the bus. A short time later, a fireball erupted and, again based on witnesses' accounts, it would seem that a flashover had occurred.

Four students and one instructor were in the bus when the flashover occurred. Two students and the instructor made it out under their own power, but two students remained trapped inside as the front passenger door was jammed and would not open. Crews on the outside began extinguishment and rescue efforts. Eventually the two trapped students were removed. One of the trapped students stopped breathing after being rescued, but was revived on the scene. All of the students involved had second and third degree burns to various parts of their bodies. Each remained hospitalized, the worst until February 1997.

Could this tragic event been averted? The answer is a resounding "Yes." The investigation into this incident found fifteen deviations from acceptable practices. Included in the findings were an inappropriate facility, no incident management system, no safety officer, no written plan for the exercise, no preburn tour, inadequate number of instructors for students, no water supply, no backup hose lines, no roof ventilation points, inadequate use of protective equipment, and the lack of emergency medical services on site. Unfortunately, many of these deviations have occurred before in fire service training exercises.

Why, then, did an accident like this occur again? Because the department failed to follow established accepted national standards. The National Fire Protection Association's standard *Live Fire Training Evolutions in Structures* (NFPA 1403), published prior to this event, covers virtually all of the deviations noted in this incident.

INTRODUCTION

The knowledge and understanding of safety-related standards and regulations is essential in all emergency service functions. "We have not adopted that standard" or "We are not an OSHA state" are not necessarily valid reasons to disregard the existence of standards and regulations.

regulations
rules or laws promulgated at the federal, state, or local level with a requirement to comply

standards
often developed through the consensus process, standards are not mandatory unless adopted by a governmental authority

Code of Federal Regulations
the document that contains all of the federally promulgated regulations for all federal agencies

There are differences between **regulations** and **standards.** Regulations are promulgated at some level of government by a governmental agency and have the force of law. Standards, sometimes called *consensus standards,* do not have the weight of law unless they are adopted as law by an authority having jurisdiction.

From the safety program management standpoint, an understanding of these applicable regulations and standards can be invaluable. These documents can actually provide a road map for an agency's safety and health program. This chapter focuses on the regulations and standards that apply to emergency service safety and health. Emphasis is placed on those that have the greatest effect and impact on the programs, but others are introduced also.

REGULATIONS VERSUS STANDARDS

Regulations

Regulations carry the weight of law and are mandatory in their requirements based upon federal, and in some cases, state and local legislation. The entire collection of federal regulations is contained within the fifty titles of the **Code of Federal Regulations** (CFR). These mandatory requirements impact on emergency service safety and health programs, primarily those regulations found in Title 29 CFR, which is the Occupational Safety and Health Administration's (OSHA) regulations. These specific standards are discussed in further detail later in this chapter in the section Occupational Safety and Health Administration Regulations. Remember, these regulations have the power of law and are enforced by the federal agency responsible for them.

■ **Note**
These regulations have the power of law and are enforced by the federal agency responsible for them.

Some states have adopted federal laws, which then become mandatory state requirements. For example, the State of Florida Department of Labor has adopted several OSHA regulations in their entirety including 1910.120, hazardous waste operations, and 1910.1030, exposure to bloodborne pathogens. When states adopt a regulation, then state officials provide the enforcement.

Standards

consensus standards
standards developed by consensus of industry or subject area experts, which are then published and may or may not be adopted locally; even if not adopted as law, these standards can often be used as evidence for standard of care

Other published documents do not mandate compliance. These are commonly known as **consensus standards** because a group of professionals with a specific expertise have agreed on how a specific task should be performed. The National Fire Protection Association (NFPA) is one such standards-making group. The NFPA has no enforcement authority or power; its standards are considered advisory. However, a jurisdiction can adopt an NFPA standard and then the adopting authority has legal rights to enforce the standard. The most common example of this process is the adoption of NFPA 101, *Life Safety Code.* The standard has been adopted by both local and state governments as part of the fire prevention program. Once a standard is adopted, the adopting governmental agencies have the power for enforcement.

LIBRARY
THE UNIVERSITY OF TEXAS
AT BROWNSVILLE
Brownsville, Tx 78520-4991

■ Note
Once a standard is adopted, the adopting governmental agencies have the power for enforcement.

The NFPA has developed a number of standards that impact safety and health programs. The most notable is NFPA 1500, *Fire Department Occupational Safety and Health Program*. Other standards in the 1500 series include standards for medical requirements, infection control, and fire department safety officer. Further discussion of the NFPA standards is presented later in this chapter in the section NFPA Standards.

Another consensus standard-making organization is the Alliance for Fire and Emergency Management (AFEM). Although in 1996 the alliance ceased operations, the standard still exists, as do many of the suborganizations of the alliance. One such subgroup, the International Society of Fire Service Instructors, has been publishing performance standards on industrial fire protection for many years. Having seen a need for additional performance standards relating to safety and fire operations, in 1995 the AFEM published the performance standard for incident safety officer.

standard of care
the concept of what a reasonable person with similar training and equipment would do in a similar situation

It is important to remember that these consensus standards are not mandatory unless adopted into law by local or state legislation. However, because a group of professionals with related interest and expertise have agreed on some minimum level of performance, these standards tend to become a **standard of care** in the particular subject.

STANDARD OF CARE

The concept of standard of care is well known in the emergency medical field. From the first day of training, this concept is introduced to make students realize that to avoid liability they must perform in the same way as another reasonable person with the same training and equipment would perform. The concept is simple—everyone has certain expectations when it comes to performance. When a deposit is made to your checking account at the bank, you have a reasonable expectation that the teller will put the money into the right account. If mistakes are made, the bank has to take responsibility for the teller's actions.

■ Note
The concept of standard of care is simple—everyone has certain expectations when it comes to performance.

This same concept can be applied to safety and health issues and closely relates to the existence of standards. The publication of a safety standard in and of itself has an impact on standard of care. For example, NFPA 1500 requires that all responders to hazardous situations be equipped with a **personal alert safety system** (PASS). Should a firefighter get lost in a building fire and die, you may find yourself in court answering the question of why a PASS device was not provided. But NFPA is not law and has not been legally adopted in my jurisdiction you say. This answer will not be a viable defense when the firefighter's family sues for not following reasonable industry standards, because the standard of care has been defined by the NFPA document.

personal alert safety system
a device that produces a high-pitched audible alarm when the wearer becomes motionless for some period of time; useful to attract rescuers to a downed firefighter

Standard of care is not a static concept but instead is very dynamic. It changes with new technologies and the development of regulations, standards, and procedures. Twenty-five years ago it was an acceptable standard of care to allow firefighters to fight fires without PASS devices. They did not exist and, if they did, no published document required them, therefore it was an acceptable practice.

OCCUPATIONAL SAFETY AND HEALTH ADMINISTRATION REGULATIONS

A number of OSHA regulations affect emergency service safety and health programs. Those most talked about in recent years include the bloodborne pathogens rule, confined space operations, hazardous waste operations, and respiratory protection. Each of these regulations is described in this section in terms of applicability to safety and health programs and a review of the contents of each is provided.

The application of OSHA regulations to public employers is a confusing issue. Under the Occupational Safety and Health Act of 1970, federal OSHA has no direct enforcement authority over state and local governments. However, a state may opt to implement its own enforcement program and may do so as long as federal OSHA approves the state's safety and health plan. The twenty-five states and territories in the following list currently have state OSHA plans. In these states all paid firefighters, including state, county, or municipal, are covered by regulations promulgated by federal OSHA.

States and Territories with OSHA Plans

Alaska
Arizona
California
Connecticut
Hawaii
Indiana
Iowa
Kentucky
Maryland
Michigan
Minnesota
Nevada
New Mexico
New York
North Carolina
Oregon
Puerto Rico
South Carolina
Tennessee
Utah
Vermont
Virginia
Virgin Islands
Washington
Wyoming

OSHA 1910.146 (29 CFR 1910.146) Permit-Required Confined Spaces

This regulation applies to permitted confined space operations. There are several areas that apply to safety practices and procedures. This document defines a confined space as any area that (1) is large enough and so configured that an employee can bodily enter and perform assigned work; (2) has limited or restricted means for entry or exit (for example, tanks, vessels, silos, storage bins, hoppers, vaults, and pits are spaces that may have limited means of entry); and (3) is not designed for continuous employee occupancy (see Figure 2-1).

Clearly, there are cases when the emergency service employee would be required to enter the environments as just defined. Requirements contained in this document include those for written plans, atmosphere monitoring, notification, and equipment. Some companies that have permitted confined spaces on site have contracted with the local fire or rescue agency to perform services if needed. Even in the absence of this pre-event agreement, we can expect that when an emergency occurs, the local emergency services will be summoned by a 911 phone call. From a safety program standpoint, these operations should not be performed until written procedures have been developed, the necessary equipment obtained, and the personnel have been properly trained.

■ Note **From a safety program standpoint, these operations should not be performed until written procedures have been developed, the necessary equipment obtained, and the personnel properly trained.**

OSHA 1910.134 (29 CFR 1910.134) Respiratory Protection

This regulation requires that respirators, including self-contained breathing apparatus (SCBA), be provided by the employer when such equipment is necessary to protect the health of the employee. The employer is required to provide the respirators applicable and suitable for the purpose intended. The employer is responsible for the establishment and maintenance of a respiratory protection program. The employee shall use the provided respiratory protection in accordance with instructions and training received. Requirements for a minimal acceptable program are:

1. Written **standard operating procedures** shall govern the selection and use of respirators.
2. Respirators shall be selected on the basis of hazards to which the worker is exposed.
3. The user shall be instructed and trained in the proper use of respirators and their limitations.
4. Respirators shall be regularly cleaned and disinfected. Those used by more than one worker shall be thoroughly cleaned and disinfected after each use.
5. Respirators shall be stored in a convenient, clean, and sanitary location.
6. Respirators used routinely shall be inspected during cleaning. Worn or deteriorated parts shall be replaced. Self-contained respirators for emergency use, such as shown in Figure 2-2, shall be thoroughly inspected at least once a month and after each use.
7. Appropriate surveillance of work area conditions and degree of employee exposure or stress shall be maintained.
8. Regular inspections and evaluations shall be conducted to determine the continued effectiveness of the program.

standard operating procedures
sometimes called standard operating guidelines, these are department-specific operational procedures, policies, and rules made to assist with standardized actions at various situations

Figure 2-1 *Confined space entry is governed by OSHA regulations.*

Figure 2-2 *SCBA usage falls under OSHA's respiratory protection regulation.*

immediately dangerous to life and health
used by several OSHA regulations to describe a process or an event that could produce loss of life or serious injury if a responder is exposed or operates in the environment

Even from these minimum requirements one can see immediately the safety program implications. Another section of the regulation requires that the fitness of an employee be evaluated prior to being permitted to wear a respirator. This evaluation is to be performed by a licensed physician. The regulation also sets forth standards for air quality and SCBA maintenance programs.

One controversial requirement in the document is that of staffing requirements. In January 1998, OSHA issued a final formal interpretation of federal regulations and compliance orders regarding the use of SCBA. In summary, the ruling states that when workers are using SCBA in **immediately dangerous to life and health** (IDLH) environments or unknown environments, the worker shall operate in a buddy system and there shall be at least two persons outside the danger area ready to effect a rescue. The rule has come to be known as the two in–two out rule and has significant implications. The rule does not preclude the firefighters from attempting a rescue in a known life and death situation prior to the rescue team being in place. The implication of this ruling is that before firefighters can operate in situations where SCBA is worn, four firefighters are required. This ruling is significant to the operations of both career and volunteer fire service agencies. Appendix III contains an excellent and concise document issued by the International Association of Fire Chiefs (IAFC) and the International Association of Fire Fighters (IAFF) that deals with this ruling.

The requirements of this regulation should be carefully evaluated by the safety staff prior to safety program development. Procedures that address opera-

■ Note
The OSHA ruling states that when workers are using SCBA in immediately dangerous to life and health (IDLH) environments or unknown environments, the worker shall operate in a buddy system and there shall be at least two persons outside the danger area ready to effect a rescue.

tions when respirators are worn, the maintenance and repair of respirators, and the training that employees receive must be designed to meet this regulation.

OSHA 1910.120 (29 CFR 1910.120) Hazardous Waste Operations and Emergency Response (HAZWOPER)

The emergency service application of this regulation is clearly described as emergency response operations for releases of, or substantial threats of releases of, hazardous substances without regard to the location of the hazard. This OSHA regulation has a number of health and safety related requirements including a written safety and health program for employees involved in hazardous waste operations. This written plan must be made available to all employees and is required to contain, as a minimum the following sections: (1) a means to identify, evaluate, and control safety and health hazards and provide for emergency response for hazardous waste operations; (2) an organizational structure; (3) a comprehensive workplan; (4) a site-specific safety and health plan, which need not repeat the employer's standard operating procedures; (5) a safety and health training program; (6) a medical surveillance program; (7) the employer's standard operating procedures for safety and health; and (8) any necessary interface between general program and site-specific activities.

The regulation also has specific requirements for personal protective equipment, training levels, and the necessary personnel to operate at an applicable incident.

For the safety program in a department that has the opportunity to respond to hazardous material incidents, familiarization and compliance with this regulation is necessary. The 1910.120 regulation was referenced in the Jan 1998 OSHA interpretation on staffing, specifically the requirement of personnel to utilize a buddy system (see Figure 2-3).

OSHA 1910.156 (29 CFR 1910.156) Fire Brigades

fire brigades
trained personnel within a business or at an industrial site for fire fighting and emergency response

The OSHA fire brigade regulation applies to **fire brigades,** industrial fire departments, and private or contractual type fire departments. The personal protective clothing requirements of this regulation apply only to those fire brigades that perform interior structural firefighting. The regulation specifically excludes airport crash fire rescue and forest firefighting operations. This regulation contains requirements for the organization, training, and personal protective equipment for fire brigades whenever they are established by employers.

The fire brigade regulation contains sections on (1) organizational statement (which establishes the existence of the brigade and personnel requirement), (2) training and education, (3) firefighting equipment, (4) protective clothing, and (5) respiratory protection.

As with the 1910.134 and 1910.120 regulations, this regulation was also referenced in the Jan 1998 interpretation on staffing. Specifically section (f)(1)(i) was cited as it requires that SCBA be provided for all firefighters engaged in interior structural firefighting.

OSHA 1910.1030 (29 CFR 1910.1030) Occupational Exposure to Bloodborne Pathogens

The bloodborne pathogens regulation became effective on March 6, 1992. The intent of the regulation was to eliminate or minimize occupational exposures to hepatitis B virus (HBV), human immunodeficiency virus (HIV), and other bloodborne pathogens. The regulation is based on the premise that these exposures can be minimized or eliminated through a combination of engineering and work practice controls, personal protective clothing and equipment, training, medical surveillance, hepatitis B vaccination, signs and labels, and other provisions. The regulation applies to all employees who may have occupational exposure to blood or other potentially infectious material as defined in the regulation (see Figure 2-4).

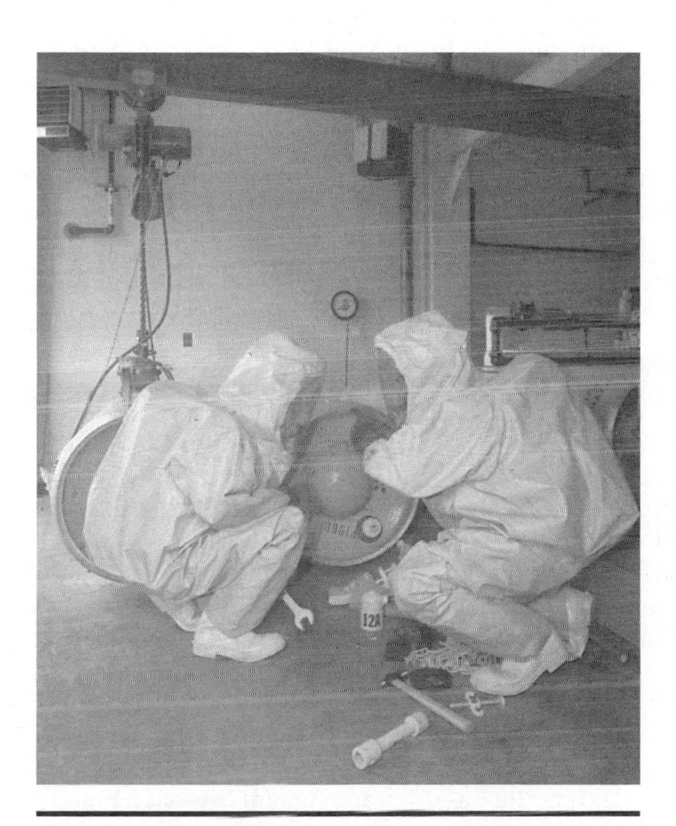

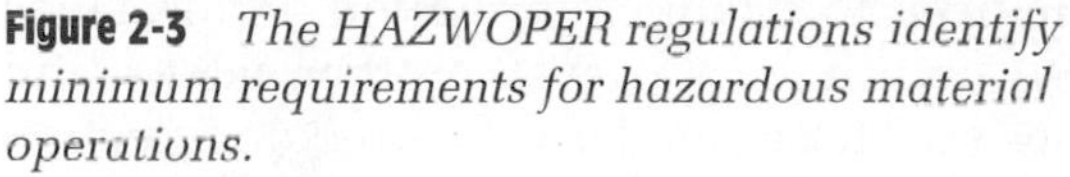

Figure 2-3 *The HAZWOPER regulations identify minimum requirements for hazardous material operations.*

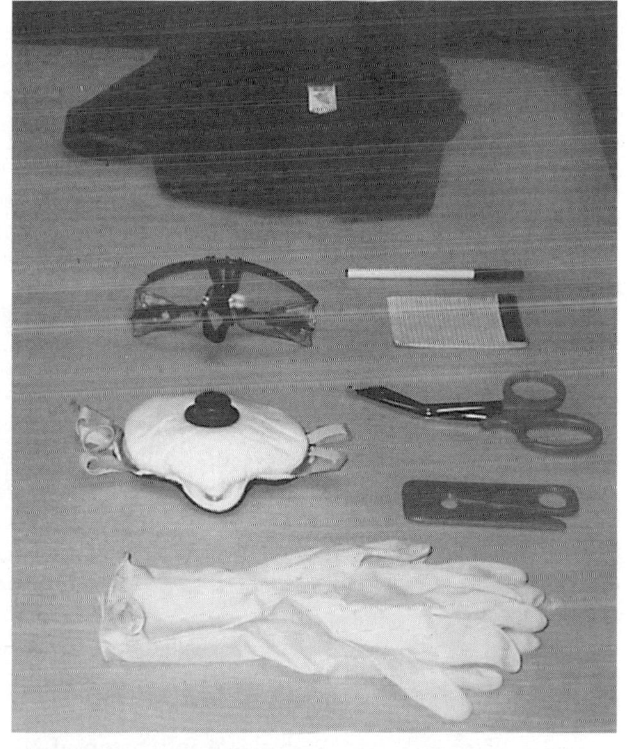

Figure 2-4 *Personnel protective equipment is required to limit exposure to bloodborne pathogens.*

■ Note
This regulation requires employers to have an exposure control plan that is reviewed and updated annually or sooner, as needed, to reflect new procedures or new equipment.

This regulation requires employers to have an exposure control plan that is reviewed and updated annually or sooner, as needed, to reflect new procedures or new equipment. Other requirements of the regulation include exposure determination, engineering and work practice controls, labeling of waste, housekeeping, personal protective clothing and equipment, record keeping, and decontamination procedures. The 1910.1030 regulation requires employers to provide hepatitis vaccinations to all employees with occupational exposure risk.

This regulation must be consulted when writing medical/rescue response procedures for the department's safety program, because almost all emergency service employees fall within the scope of this document in terms of occupational exposure.

OSHA General Duty Requirement

When the Occupational Safety and Health Act was enacted in 1970, it was clear to Congress that not all safety and health situations could be covered by the regulations. Knowing this, Congress included in the act the General Duty Clause.

The general duty clause states that each employer shall furnish to each of its employees employment and a place of employment that are free of recognized hazards that are causing or are likely to cause death or serious harm to their employees. Generally the OSHA Review Commission and court precedent have shown that the following elements would be necessary to prove a violation of the general duty clause:

■ Note
In the absence of a specific OSHA regulation, a national consensus standard (such as NFPA 1500) can be used as reference for guidance in enforcement of the general duty clause.

- The employer failed to keep the workplace free of a hazard to which employees of the employer were exposed.
- The hazard was recognized.
- The hazard was causing or was likely to cause death or serious injury.
- There was a feasible and useful method to correct the hazard.

The implication of this clause to the safety and health program comes from a stated relationship with national consensus standards such as the NFPA. In the absence of a specific OSHA regulation, a national consensus standard (such as NFPA 1500) can be used as reference for guidance in enforcement of the general duty clause. This is an important consideration for the safety and health program manager when assessing the need to comply with consensus standards.

NFPA STANDARDS

With the publication of NFPA 1500 in 1987 and the movement toward a greater concern for firefighter safety came the need to establish minimum performance criteria. To fulfill this need the NFPA continued to develop other safety- and health-related standards in the 1500 series. These standards focused on infection control, fire department safety officers, medical requirements, and incident management systems. Because each of these has an impact on safety program management, each is described in this section.

NFPA 1500: *Fire Department Occupational Safety and Health Program*

Clearly the most significant document in the development of the firefighter safety movement was the adoption of NFPA 1500, after having the most public comments in the history of the NFPA's standards-making process. This standard was the first consensus standard for fire service occupation safety and health programs. It was thought that this standard would financially break fire departments and possibly put some out of business completely. In 1992, the second edition was adopted amid controversy surrounding safe fireground staffing. In 1997, the third edition of 1500 was adopted.

As in previous editions, the 1997 edition of NFPA 1500 covers the essential elements of a comprehensive safety and health program. This document also incorporates other standards into it as requirements: NFPA 1500 lists forty-five other documents that are incorporated by reference, thirty-five of which are other NFPA standards (see Figure 2-5).

The NFPA standard 1500 provides the framework for any fire service-related safety and health program, including considerations for emergency medical services (EMS) response. The requirements outlined in 1500 are considered the minimum for an organization providing rescue, fire suppression, and other emergency service functions whether full- or part-time, as public, governmental, military, private, and industrial fire departments. The standard is not intended to apply to industrial fire brigades, which are covered in NFPA 600. Industrial fire brigades differ from industrial fire departments in that members of industrial fire departments have the primary responsibility for fire protection, whereas the primary responsibility of industrial fire brigade members is elsewhere in the facility, and they only handle fires as they first occur or are in the incipient stage. Again, keep in mind that the standard is a minimum standard; nothing restricts the program manager from going beyond these requirements.

The many facets of the safety program covered by NFPA 1500 are administration, organization, risk management, training and education, vehicle and equipment, protective clothing and protective equipment, emergency operations, facility safety, medical and physical requirements, and member assistance programs. Appendix A of NFPA 1500 is not considered part of the standards requirements, but provides additional information about selected requirements. Appendix B lists referenced documents for informational purposes but they are not considered requirements of the document.

NFPA 1500 can provide the user with comprehensive direction for developing a safety and health program. Even if not adopted, the standard is known throughout the emergency field and is accepted as a standard of care. No safety program manager should be without a copy of 1500 and its referenced documents.

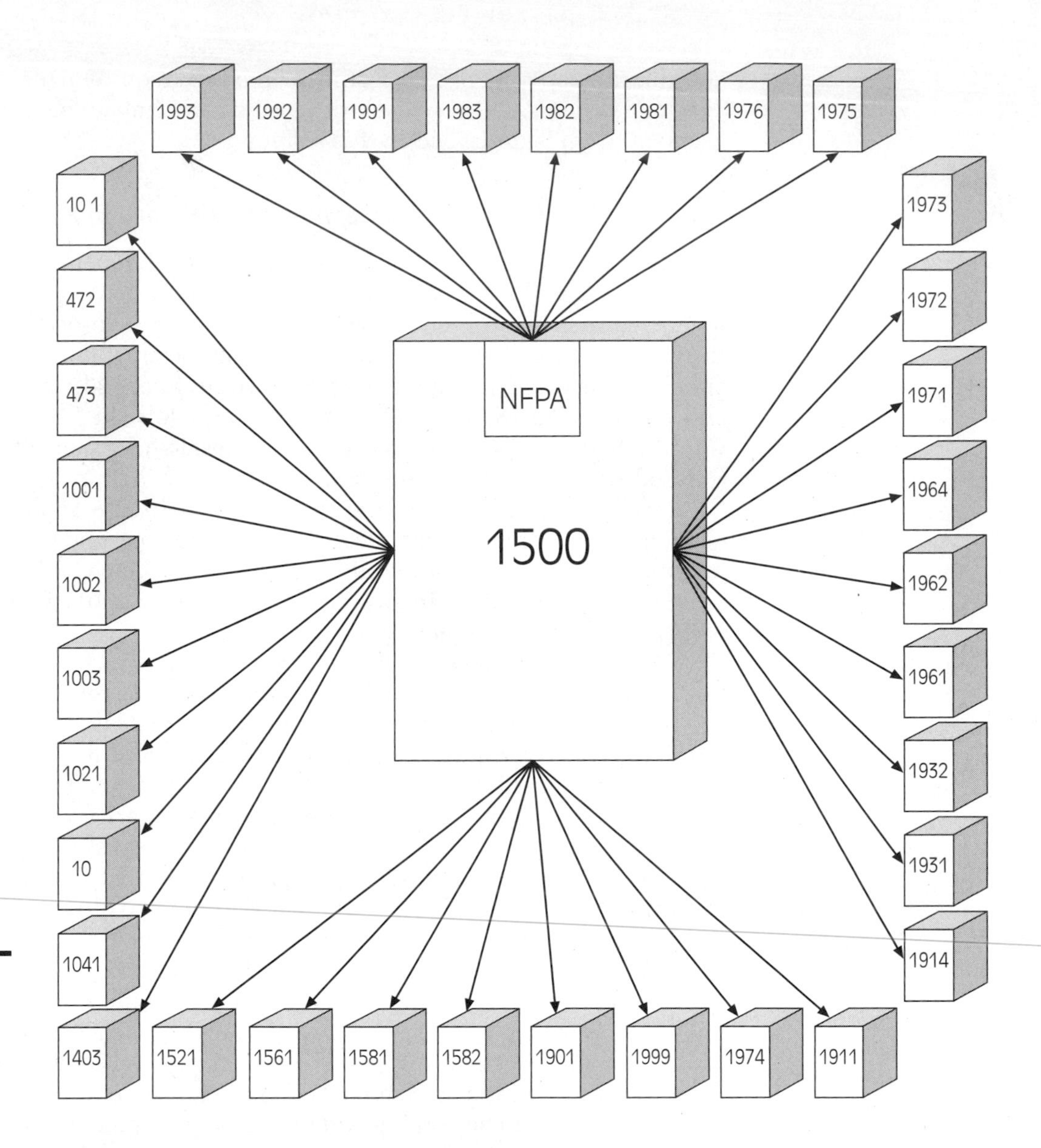

Figure 2-5 *The NFPA 1500 standard directly references 35 other NFRA standards.*

NFPA 1521: *Fire Department Safety Officer*

With the 1997 edition of *Fire Department Safety Officer,* the NFPA changed the standard number from 1501 to 1521 to better fit the numbering plan for fire service occupational safety and health documents. This standard contains the minimum requirements for the assignment, duties, and responsibilities of a fire department or other fire service safety officer (Figure 2-6).

Figure 2-6 *A safety officer should be in place throughout the incident.*

The standard provides for the assignment, qualifications, and authority of fire department safety officers. It further requires that the fire department safety officers manage the collection and analysis of all occupational-related accident, injury, and death records as required by NFPA 1500 Section 2-7.

The standard requires that the safety officer be a member of the department's safety and health committee and be the liaison with the fire chief on safety-related matters and recommendations. Other sections contained in this standard include the safety officer's responsibility for establishing rules, regulations, and procedures relating to the safety and health program, the role in an accident prevention program, review of specifications for new apparatus and equipment, accident investigation, incident scene safety, and training and education.

NFPA 1561: *Fire Department Incident Management System*

At the time of the adoption of NFPA 1500, which requires that fire scene operations be conducted using an incident management system, many incident management systems (IMS) were being used. One concern was that many of these systems did not consider employee safety and health as a major concern. In addition to requiring an IMS, the NFPA committee determined that there should be some specific performance criteria for the components of the system. Hence the development of standard 1561, which contains the minimum requirements for an IMS that can be used for all emergency incidents. The standard describes the system structure including implementation, interagency coordination, command structure, training, and qualifications. Chapter 3 of NFPA 1561 describes the system components, the incident commander, command staff, planning functions, logistic functions, operations functions, communications, staging and finance functions. Chapter 4 contains the role and responsibilities of the incident command and supervisory personnel and the requirements for personnel accountability, rest, and rehabilitation.

The safety program manager will find this document very helpful when developing procedures and training for incident command.

NFPA 1581: *Fire Department Infection Control*

Fire Department Infection Control, standard 1581, was developed to reduce the exposures to infectious diseases by responders in both emergency and nonemergency situations. The standard is compatible with the guidelines developed by the Centers for Disease Control described later in this chapter under Other Related Standards and Regulations. The standard defines the minimum requirements for fire department infection control programs, including elements for reducing exposure in both emergency and nonemergency settings. The standard begins with a description and requirements for the program components including, policy, training and education, infection control liaison person, immunization and testing, and exposures. Chapter 3 of the standard sets forth the requirements for the department's facilities such as disinfecting areas, cleaning areas, and storage rooms. Chapter 4 describes requirements related to emergency medical operations including personnel issues, infection control garments and equipment, and the handling of sharp objects. Skin washing, disinfectants, emergency medical equipment, clothing, and disposal of material are addressed in Chapter 5 of the standard. This standard is of value as a resource to assist in compliance with the OSHA 1910.1030 regulation.

NFPA 1582: *Medical Requirements for Fire Fighters*

Standard 1582 was developed to cover the medical requirements necessary for persons engaged in fire-fighting activities. The requirements in this standard are intended to apply to both existing employees and to employment candidates. There are two categories of medical conditions in the standard, A and B. Persons with category A conditions are not allowed to perform fire-fighting functions. Category B conditions are to be evaluated on a case-by-case basis with the fire department physician. This standard contains the medical requirements for firefighters whether full or part time, paid or volunteer, and specifies the minimum medical requirements.

The medical process is described in Chapter 2 of the Standards and includes the requirements for the medical evaluation, fire department physician, preplacement medical evaluation, periodic medical evaluation, and return-to-duty medical evaluation. This standard is helpful in determining the frequency of physical exams, developing criteria for the physical exam, and provides direction to a department physician as to what conditions may disqualify firefighters from performing their duties.

Other NFPA Standards

There are thirty-five NFPA standards incorporated by reference into NFPA 1500. Four of these standards are from the 1500 series and were described previously. The other thirty-one are described in the following list of standards that affect safety and health from other standpoints, for example, apparatus specifications, live fire training, and performance requirements for protective clothing. It is important to at least recognize the existence and inclusion of these standards in NFPA 1500. The following list describes these thirty-one standards and others that apply in part to safety and health.

NFPA 10, *Portable Fire Extinguishers,* applies to the selection, installation, inspection, maintenance, and testing of portable extinguishing equipment.

NFPA 101, *Life Safety Code,* identifies the minimum life safety requirements for various occupancies.

NFPA 472, *Professional Competence of Responders to Hazardous Materials Incidents,* identifies the competencies for first responders at the awareness level, first responders at the operational level, hazardous materials technicians, incident commanders, and off-site specialist employees.

NFPA 473, *Competencies for EMS Personnel Responding to Hazardous Materials Incidents,* identifies the requirements for basic and advanced life support personnel in the prehospital setting.

NFPA 1001, *Fire Fighter Professional Qualifications,* identifies performance requirements, specifically the minimum requirements for firefighter candidates and the two levels thereafter.

NFPA 1002, *Fire Department Vehicle Driver/Operator Professional Qualifications,* identifies minimum job performance for the firefighter driver/operator of fire department vehicles.

NFPA 1003, *Airport Fire Fighter Professional Qualifications,* identifies the minimum job performance for the airport firefighter.

NFPA 1021, *Fire Officer Professional Qualifications,* identifies the requirements necessary to perform the duties of a fire officer and establishes four levels of progression.

NFPA 1041, *Fire Service Instructor,* identifies the professional levels of competence required for fire service instructors.

NFPA 1403, *Live Fire Training Evolutions in Structures,* requires the establishment of procedures for training of fire suppression personnel engaged in structural firefighting under live fire conditions (see Figure 2-7).

NFPA 1901, *Fire Department Apparatus,* applies to all new automotive fire apparatus. The 1996 edition has combined NFPA 1902, 1903, and 1904 into one standard to cover all types of fire department apparatus as opposed to having a different standard for each.

NFPA 1911, *Service Tests of Pumps on Fire Department Apparatus,* applies to the service testing of fire pumps and attack pumps of fire department automotive apparatus excluding apparatus equipped solely with pumps rated at less than 250 gpm (950 L/min).

NFPA 1914, *Testing Fire Department Aerial Devices,* applies to the inspection and testing of all fire apparatus.

NFPA 1931, *Design of and Design Verification Tests for Fire Department Ground Ladders,* applies to the specific requirements for

Figure 2-7 *NFPA 1403 defines minimum requirements for live fire training exercises.*

the design of and the design verification test for all new fire department ground ladders intended for use by fire department personnel for firefighting operations and training.

NFPA 1932, *Use, Maintenance, and Service Testing of Fire Department Ground Ladders,* applies to the specific requirements for the use, maintenance, inspection, and service testing of fire department ground ladders.

NFPA 1961, *Fire Hose,* applies to the testing of new fire hose, specified as attack hose, occupant use hose, forestry hose, and supply hose.

NFPA 1962, *Care, Use, and Service Testing of Fire Hose Including Couplings and Nozzles,* applies to the care of all types of fire hose, coupling assemblies, and nozzles while in service, in use, and after use.

NFPA 1964, *Spray Nozzles (Shutoff and Tip),* applies to portable adjustable pattern nozzles intended for general fire department use.

NFPA 1971, *Protective Clothing for Structural Fire Fighting,* specifies criteria and test methods for protective clothing designed to protect firefighters against environmental effects during structural firefighting.

NFPA 1972, *Helmets for Structural Fire Fighting,* specifies criteria and test methods for helmets for structural firefighting designed to mitigate environmental effects to the firefighter's head.

NFPA 1973, *Gloves for Structural Fire Fighting,* specifies design, performance, and test methods for gloves designed to protect firefighters against adverse environmental effects to the hands and wrists during structural firefighting. This standard also specifies performance and test methods designed to protect firefighters against skin exposure to blood or other liquid-borne pathogens that might exist in an emergency.

NFPA 1974, *Protective Footwear for Structural Fire Fighting,* establishes design, performance, and test methods for protective footwear designed to mitigate environmental effects to the foot and ankle during structural firefighting (see Figure 2-8).

NFPA 1975, *Station/Work Uniforms for Fire Fighters,* specifies general requirements, performance requirements, and test methods for materials used in the construction of station/work uniforms to be worn by members of the fire service.

NFPA 1976, *Protective Clothing for Proximity Fire Fighting,* specifies design, performance, and test methods for protective clothing designed to provide limb/torso projections for firefighters.

NFPA 1981, *Open-Circuit Self-Contained Breathing Apparatus (SCBA) for Fire Fighters,* specifies requirements for design, performance, testing, and certification of SCBA used in firefighting, rescue, and other hazardous situations.

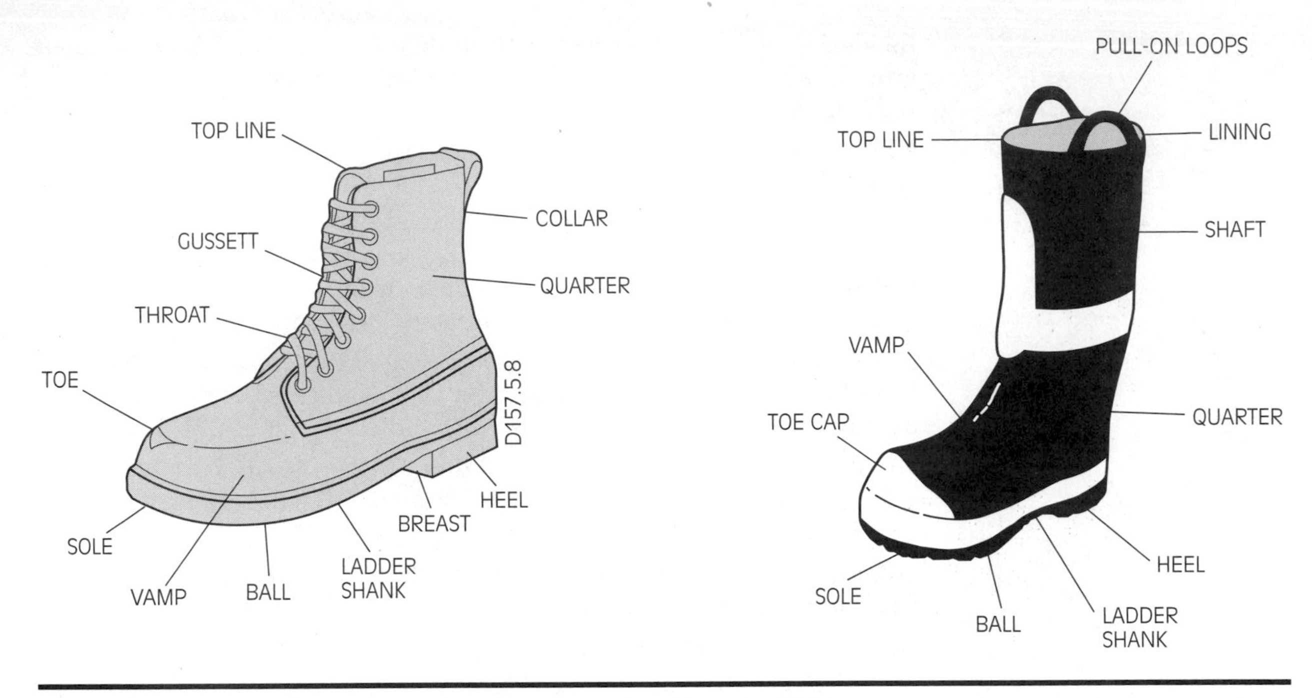

Figure 2-8 *Two types of NFPA-compliant foot protection. Courtesy USFA.*

NFPA 1982, *Personal Alert Safety Systems (PASS) for Fire Fighters,* specifies design, performance, and test methods for personal alert safety systems to be used by firefighters engaged in firefighting, rescue, and other hazardous duties.

NFPA 1983, *Fire Service Life Safety Rope, Harness, and Hardware,* specifies performance, design, and test methods for life safety rope, harness, and hardware used by the fire service.

NFPA 1991, *Vapor-Protective Suits for Hazardous Chemical Emergencies,* applies to the requirements for vapor-protective suits and replacement components.

NFPA 1992, *Liquid Splash-Protective Suits for Hazardous Chemical Emergencies,* applies to the requirements for liquid splash-protective suits and replacement components.

NFPA 1993, *Support Function Protective Clothing for Hazardous Chemical Operations,* applies to the design, performance, and test methods for protective clothing used by personnel in support functions during hazardous chemical operations.

NFPA 1999, *Protective Clothing for Emergency Medical Operations,* specifies minimum emergency medical clothing designed to protect emergency personnel or patients from exposure to liquid-borne pathogens during emergency medical operations.

The following three standards are not specifically referenced by NFPA 1500 but also impact an organization's safety and health program and therefore deserve review:

NFPA 1404, *Fire Department Self-Contained Breathing Apparatus Program,* contains the minimum requirements for a fire service respiratory protection program.

NFPA 1410, *A Training Standard on Initial Fire Attack,* addresses the evaluation of prior training in initial fire flow delivery procedures by fire department personnel engaged in structural firefighting.

NFPA 1470, *Search and Rescue Training for Structural Collapse Incidents,* identifies and establishes levels of training for safely and effectively conducting operations at structural collapse incidents.

ALLIANCE FOR FIRE AND EMERGENCY MANAGEMENT STANDARD 502

The Alliance for Fire and Emergency Management was an umbrella organization for several fire service organizations including the International Society of Fire Service Instructors, the Fire Department Safety Officers Association, and the Fire Officers Association. Before the alliance ceased operations in 1996, it published a performance standard for incident scene officers. Although the parent or umbrella organization is no long doing business, the other organizations are, therefore a discussion of the standard is warranted.

Alliance standard 502, *Incident Scene Safety Officer,* was developed after research by the Fire Department Safety Officers Association revealed a need to develop professional competencies for the fire department safety officers. The Professional Qualification Council for Fire Department Safety was convened. This council felt that four basic performance levels would be needed to fulfill the certification requirements of fire department safety officers: (1) firefighter/basic safety, (2) incident scene safety officer, (3) fire department safety administrator, and (4) incident scene safety officer-special operations. The Professional Qualifications Council decided that the current NFPA Standard 1521 for fire department safety officers did not adequately address the performance levels identified. Therefore the council requested that a performance standard be developed for each of the afore-mentioned levels. The priority for the first standard was for incident scene safety officer.

The 502 standard contains the minimum performance requirements for the individual who is to perform as the safety officer at routine incidents. The standard applies to all fire departments, organizations, and agencies, public or private, paid or voluntary, that provide fire suppression, rescue, emergency medical, or other services to manage emergencies. Further the standard is intended to provide equivalency for NFPA 1521, 1992 edition, as allowed in paragraph 1-3.

This standard contains thirty-one chapters dealing with a multitude of incidents and hazards associated with operations at these incidents and with various tools and equipment. Each chapter includes broad performance requirements for the specific task or equipment. Included are pneumatic, hydraulic, and mechanical tool hazards; electrical, ergonomic, pressure vessel, and boiling liquid expanding vapor explosion (BLEVE) hazards; hazardous materials; slip and fall hazards; violence awareness; accidents and close call events; personnel accountability; incident management systems; training safety; chemical, biological, and radiological hazards; thermal stress; sound hazards; psychological hazards; personal protective envelope; infection control; vehicle, extrication, fireground, unique, forcible entry, and search and rescue operations.

Like the NFPA 1500 series of standards, this standard should be part of the safety program manager's library. As a voluntary standard it cannot be enforced unless adopted by local government, but could be referred to as a standard of care or used under OSHA's General Duty Clause. For the incident safety officer this standard is both a valuable reference and a practical guide to responsibilities.

SOURCES OF OTHER RELATED STANDARDS AND REGULATIONS

American National Standards Institute

The American National Standards Institute (ANSI) is a national voluntary standard-making organization that uses the consensus process. Its membership is comprised of representatives from many disciplines with expertise in the area of the standard being developed. These members may include professional, trade, technical, or consumer organizations, state and federal agencies, and individual companies.

Many of the standards developed by ANSI are in response to requirements in other regulations and standards. For example, ANSI has a standard entitled *Protective Clothing for Emergency Medical Operations,* which matches NFPA 1999, 1992 edition. Other approved consensus standards may become ANSI standards. For example, some standards developed by the NFPA also are approved by the ANSI Safety and Health Standards Board.

American Society of Testing and Materials

The American Society of Testing and Materials (ASTM) is a not-for-profit organization that develops standards for materials, products, and services. The ASTM also is a consensus organization comprised of individuals and organizations from various disciplines. While much of the ASTM's purpose is directed at consumer protection issues, some standards impact emergency service safety. The ASTM committee F-30 has generated two standards directed toward emergency medical systems: the *Guide for Structures and Responsibilities for Emergency Medical*

Services Systems Organizations and the *Practice for Training the Emergency Medical Technician (Basic).*

Environmental Protection Agency

The Environmental Protection Agency (EPA) has promulgated standard 40 CFR 311 that mirrors the OSHA Hazardous Waste Operations and Emergency Response document. The implication of this action is most important for organizations that do not have state OSHA plans. In these states the requirements of OSHA 1910.120 apply to public employers because of this EPA regulation. Therefore a public emergency service organization in a non-OSHA state that responds to, or has the potential to respond to, hazardous substance releases must comply with the OSHA requirements.

Centers for Disease Control and Prevention

Guidelines for the Prevention of Transmission of Human Immunodeficiency Virus and Hepatitis B Virus to Health-Care and Public-Safety Workers was released by the Centers for Disease Control and Prevention in February 1989. This document provides an overview of the modes of transmission of human immunodeficiency virus (HIV) in the workplace, an assessment of the risk of transmission under various assumptions, principles underlying the control of risk, and specific risk-control recommendations for employers and workers. The document also includes information on medical management of persons who have sustained an exposure in the workplace to these viruses. This document is very useful for a department developing the exposure control plan required by OSHA 1910.1030.

Ryan White Comprehensive AIDS Resources Emergency Act of 1990, Subtitle B

Another document that must be reviewed when managing infection control is the Ryan White Act. This act sets forth the requirements for employers to designate an infection control officer. Further, the act provides for the notification of employees by medical facilities after an exposure takes place. The safety program manager must be aware of the act as the notification procedures, the designated officer, and the followup procedures should become part of the department's program.

Summary

Many regulations and standards from a number of sources relate to emergency service occupational safety and health. Regulations are law and can be enforced by the agency that promulgated them. Standards, however, are developed by consensus of experts in the subject area and only become enforceable if adopted by a state or local government. Although standards may not be officially adopted, they may be used to establish standard of care or may be referenced by a regulation or used during a legal battle.

Many applicable safety and health regulations come from OSHA. These regulations cover public employees in the twenty-five states and territories that have state plans. In other states these regulations do not apply to public agencies. OSHA has regulations for response to hazardous materials incidents, confined space rescue, respiratory protection, and fire brigades. Within the Occupational Safety and Health Act is the General Duty Clause which has implications for operations not explicitly covered by other OSHA regulations.

Consensus standards provide a guide for safety and health programs. The NFPA publishes a number of specific standards relating to firefighter safety and health. These safety standards are found in the NFPA 1500 series. Other NFPA standards are referenced within the 1500 series and therefore become part of them by reference.

Other regulations and standards exist that apply to safety and health. Included is guidelines from the Centers for Disease Control and Prevention on bloodborne diseases, American National Standards Institute standards, American Society of Testing and Materials, the Environmental Protection Agency regulations, and the Ryan White Law.

In order to develop a comprehensive safety and health program these regulations and standards must be consulted. When law they must be complied with, but if not law they can provide a framework for the safety program that is accepted and supported by national standards and trends.

Concluding Thought: Safety and health are ongoing considerations for which there are many laws and standards.

Review Questions

1. The fire chief has asked you to sit on a committee to update standard operating procedures (SOPs). Consider each of the following SOP subjects and list what regulations or standards you would refer to as you formulated the updated SOPs.

A. Confined space rescue
B. Hazardous materials response and operations
C. Infection control
D. Incident management system
E. Self-contained breathing apparatus maintenance and usage
F. Operations at single family house fires

2. Discuss the meaning of standard of care.
3. Which of the following states do not have state OSHA plans?
   A. Florida
   B. Maryland
   C. California
   D. Kentucky
4. Compare regulations to standards.
5. What three OSHA regulations were recently cited in an interpretation on fireground staffing?
6. The guidelines published by the Centers for Disease Control would help the safety program manager in complying with which OSHA regulation?
   A. 1910.120
   B. 1910.156
   C. 1910.134
   D. 1910.1030
7. In effect, 40 CFR 311 mirrors OSHA 1910.120.
   A. True
   B. False
8. The Alliance for Fire and Emergency Management Standard 502 is a consensus standard.
   A. True
   B. False
9. NFPA 1500 is intended to apply to *all* fire departments including industry fire brigades.
   A. True
   B. False
10. You are the lead instructor in a training class that is going to conduct a live burn. Which of the following NFPA standards would be helpful in planning a safe drill?
   A. 1500
   B. 1403
   C. 1410
   D. all of the above

Activities

1. Review your department's purchasing process in terms of equipment and apparatus. Does your department follow NFPA recommendations.
2. Review your department's procedures for response to confined space and hazardous materials incidents. Are your procedures in compliance with those in the respective regulations?
3. Review NFPA 1500, then study your department. Determine where your department stands in terms of compliance.

Risk Management

Learning Objectives

Upon completion of this chapter, you should be able to:

- Discuss the meaning of and the process for risk identification.
- Discuss the meaning of and the process for risk evaluation.
- Discuss the meaning of and the process for risk control.
- Apply risk control strategies to an injury problem.

Case Review[1]

The lightweight wood truss roof of a church in Memphis, Tennessee, collapsed on December 26, 1992, just 7 minutes after the first units arrived on the scene of the midafternoon arson fire. Two Memphis firefighters later died from burns that resulted from being trapped under the burning structure.

The analysis of this incident revealed a number of important factors including the short time that a lightweight wood truss roof can be expected to maintain its structural integrity when involved in fire and the lack of warning indicators of pending collapse. Firefighters must identify buildings with lightweight wood truss roof systems and know the proper tactics and strategy to employ when a fire involves this type of construction. Additional topics to be considered include the use of incident management systems, particularly with regard to operational safety and crew accountability, and the protection afforded by protective clothing systems.

The key issues in this case included:

- Lightweight wood truss roof collapsed early and without warning.
- Burns could have been reduced with fire resistant uniforms, hoods, and turnout pants.
- Members were not equipped with personal alert safety systems (PASS) that may have allowed for rescue earlier.
- Prefire planning is needed to identify buildings with lightweight wood trusses.

[1] Printed in part from the United States Fire Administration's Technical Report, "Wood Truss Roof Collapse Claims Two Firefighters."

Note Risk management, required as part of NFPA 1500, includes three components: risk identification, risk evaluation, and risk control.

INTRODUCTION

After the safety and health manager has identified the safety and health issues in the local organization and has reviewed the applicable standards, risk management must be undertaken. Risk management, required as part of National Fire Protection Association (NFPA) standard 1500, includes three components: risk identification, risk evaluation, and risk control. In this chapter each of these three concepts is presented as well as strategies to undertake to complete the process.

RISK IDENTIFICATION

risks the resultant outcome of exposure to a hazard

Risks could be defined as anything bad that could happen to an organization. A risk might be associated with a training exercise, an emergency medical incident, or simply a fall in the station. Identifying risks is complex, as in some cases you are trying to predict what could happen. Identification should include information from a number of sources including local and national resources.

Local Experience

The local experience can be reviewed in terms of historical incident data and data to help project future risks. Historical data can provide valuable information

Figure 3-1
Firefighters often have to work above ground under less-than-ideal conditions.

regarding risk and exposure based on the local history. For example, in analyzing the risk for firefighter injury from falls, the safety manager might review the equipment and procedures used to access aboveground locations (see Figure 3-1). The severity and frequency of the risk is presented in the next section, Risk Evaluation. Another resource from local experience in the evaluation of the risk is to question the responders. Responders usually have the best knowledge of the area and have often identified certain potential risks within their response areas. Examples may be a simple risk, such as identifying an area of a highway that is prone to flood during rain storms or a problem with responding units meeting at certain intersections during a response. Both of these cases are clearly a risk in terms of vehicular accidents.

■ **Note**
Another resource from local experience in the evaluation of the risk is to question the responders.

Historical and current data are useful for determining what has happened or what could happen today. Of greater difficulty is trying to predict what may happen in the future. Identification of future risk can be facilitated by contacting the local planning department, which can provide information on future road expansion, future industrial/commercial occupancies, and other changes that may occur in the community. This information will not only help to identify future risks, but also provide time to plan for the risk. An example might be a town where the tallest building is one story. This department does an excellent job of risk identification and has developed a risk management plan for the risks that have been identified. However, the city is planning to build a multistory senior citizen residential building. This planned change will significantly change the department's risk management plan as new risks include those that are common in high-rise structures.

Identification of Trends

Maintaining currency with emerging trends can also be useful in risk identification. By reading trade journals, searching the Internet, and attending conferences and seminars, the safety manager can see emerging trends. For example, the trend for fire departments to become involved in emergency medical response clearly created additional risks; the move toward hazardous material response and control created additional the risks. The trend of building houses and apartments with lightweight truss roof construction increased the risks during interior firefighting. The move toward placing supplemental restraints, airbags, in cars increases the risk to responders to vehicle accidents.

Trend analysis can help to identify future as well as current risk. The trend of reducing staffing and increasing services provided by emergency service organizations is certain to create additional risks. Some emergency medical services (EMS) systems now provide inoculations. Should your department include this involvement in the future, getting stuck with a needle and the concern for cross contamination may be an identified future risk.

Safety Audit

The safety audit can be another way of identifying risk within an organization. The audit should be based on some type of checklist in which a reviewer goes through the organization and checks for compliance. Commonly in emergency services safety programs, the NFPA 1500 standard is used as a benchmark. Deficiencies are noted and any associated risks identified. For example, the 1500 standard requires that all newly purchased protective gloves comply with the current NFPA standard on fire service gloves. If the safety audit reveals that half of the department's gloves have not been purchased under the new standard, the changes in the new standard should be reviewed against the quality of the existing gloves and the potential risk of hand injury should be identified. Safety audits can be performed in-house by safety staff or contracted to an outside safety consultant.

■ Note

One of the best ways to identify risk in terms of safety and health is to review past statistics.

Reviewing Previous Injury Experience

Clearly one of the best ways to identify risk in terms of safety and health is to review past statistics. As described in Chapter 1, it is also important to compare the local statistics to those nationwide and to form an understanding of why they may be different. The review of previous injuries complements the risk identification process basically because, if it happened and it caused an injury, it must be a risk. The one drawback to reviewing history is that you are doing just that. Just because something has happened, it may never happen again or just because something has not happened, does not mean it will not.

Note Categories help the safety manager to later sort the risks.

Once risks are identified through the four ways just described, it is useful to categorize the risk. Categories might include responding, in station, training, fire or EMS emergency, natural or man-made disasters, and hazardous materials incidents. Categories help the safety manager to later sort the risks.

RISK EVALUATION

Once the risks are identified, they should be recorded for future reference as the next step in the process is to evaluate the risks. For risk evaluation, we introduce two risk management terms, frequency and severity. Risk evaluation is done from these two perspectives.

Frequency

frequency how often a risk occurs or is expected to occur

Frequency is simply how often a particular risk is likely to occur, be it a fall or auto collision. The safety manager must evaluate how often a risk will occur in order to better assign the risk a priority later. Unfortunately there is no hard-and-fast rule for what an acceptable frequency is. It is a somewhat subjective measure that is very dependent on local conditions and the person making the evaluation. For this reason, numerical measures are not assigned to frequency, but instead just a high, medium, or low.

Note The safety manager must evaluate how often a risk will occur in order to better assign the risk a priority later.

For example, suppose our department, on any given day, has 100 emergency vehicles on the highways responding to incidents. Using historical data we determine that there are fifty traffic accidents per year, about one a week, involving the emergency vehicles. Is this frequency high, medium, or low? What about a department with twenty-five emergency vehicles and the same accident rate? Sounds high? So maybe in the smaller department the risk would get a high frequency rating. However, what if the first department had just spent a year training drivers in modified responses to nonemergency alarms and installed intersection control lights to prevent accidents? This fifty then would be considered a high frequency for them as well.

One way to determine if a frequency is high, medium, or low is to benchmark the local experience to that nationwide or to that of similar size organizations. For example, if the safety manager for the smaller department selected similar-sized departments with similar demographics and a similar number of vehicles on the road and found out the average number of accidents per year was 100, then the safety manager may rate this frequency as low.

Severity

severity
how severe the result is when a risk occurs

Severity is a measurement of how great the loss or the consequences of the loss will be. There are two variables in the measurement of severity, but again some is in the judgment of the person measuring. As with frequency, a measurement of high, medium, and low is used as opposed to a numerical rating. The following factors must also be considered when determining the severity element:

Costs Clearly the costs of a risk are measurable, however, cost can be both direct and indirect. Direct cost might include the costs of medical treatment, the overtime paid to cover a vacancy on a crew, or the cost of replacing equipment. Indirect costs are more difficult to measure, such as the loss of productivity, the loss of using the equipment, stress-related concern of coworker, morale, and possibly the cost of replacing the employee.

Organizational Impact How would the risk impact the organization? For example, the collision of a department's only ambulance would put that department out of the EMS transport business until a replacement could be put into service. Smaller departments could be crippled by an incident where multiple employees are injured, such as a chemical exposure. In a 1947 Texas City, Texas explosion, a shipboard fire and explosion involving ammonium nitrate killed almost all of the fire department personnel. Also included in organizational impact is the time it would take to recover, which might be the time it takes to rehire or retrain personnel or replace an emergency vehicle.

Frequency and Severity Together

When the program manager combines frequency and severity, the result can help to determine priorities to which risk control measures can be applied. It is common and necessary to prioritize the risks as seldom are time, money, and other required resources sufficient to control all of them. In prioritizing, it is important to remember that these two measures are somewhat subjective and very organizationally specific. However they can be quite useful.

■ Note
In prioritizing, it is important to remember that these two measures are somewhat subjective and very organizationally specific.

Let us return to our example of the town that is about to get a high-rise building for senior citizens. For simplicity, let us say that we have determined three risks associated with this change. The risks identified are: collisions while responding, trapped during firefighting, and sprains and strains associated with lifting during medical responses. A quick table can be put together (see Table 3-1) and an analysis made.

These ratings may have been given after significant study. For example, the building is to be built across from the EMS station, therefore the risk of a vehicle accident en route would be low. However, if an accident were to occur, the

Table 3-1 *Frequency and severity of identified risks.*

Risk	Frequency	Severity
Collisions while responding	L	H
Trapped during firefighting	L	H
Sprains and strains associated with lifting during medical responses	H	M

severity would be high because the department only has one ambulance. The building is being built to current fire codes and has a state of the art private fire protection and detection system. Although the severity of becoming trapped in a high-rise is high, the potential for this to occur in this building is low. Finally, the safety manager in this case called a fire/EMS department in another town that has a similar building. That department reported that it responds to a minimum of one EMS incident per day at their senior high-rise and usually transports someone. The safety manager further researched his department statistics and found that the department had a higher-than-average back injury problem. Therefore, in the high-rise case, the priority risk should be the prevention of back injuries. This is not to say that the other risks are not important, only to provide a guideline for prioritization.

RISK CONTROL

risk control
common approach to risk management where measures and processes are implemented to help control the number and the severity of losses or consequences of risk to the organization

Once the risk identification and evaluation process has been completed, the safety manager must develop ways to control the risk. **Risk control** can be divided into three broad categories: risk avoidance, risk control, or risk transfer.

Risk Avoidance

To avoid risk is simply not to do the task with which the risk is associated, but this option is not usually available to the emergency response profession. By the nature of the profession, we are called when things have already gone wrong. Many times it is not possible to avoid the risk. For example, the risk of steam burns from fighting an interior fire could be completely eliminated if the fire department chose not to enter burning dwellings. If the risk of exposure to infectious diseases were a priority, we could avoid the risk by not responding to calls involving blood products. But it would be difficult to justify our existence if we chose this avoidance route and did not respond to these types of incidents.

However, there are some risks that can be dealt with successfully by avoiding the risk. For example, a risk associated with starting an intravenous line (IV)

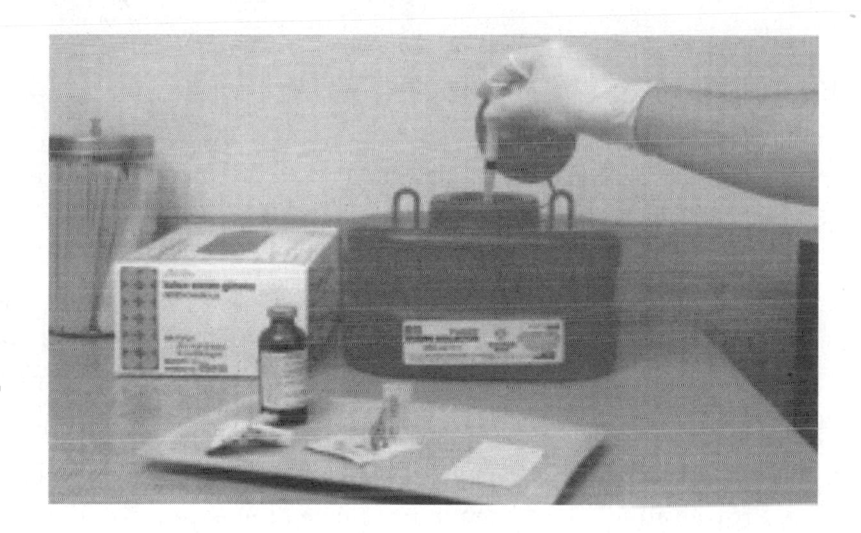

Figure 3-2 *Sharps boxes are used to eliminate needle stick injuries.*

is getting stuck with the needle when trying to resheath it. So to avoid the risk, we develop a policy that sets the procedure for dirty needles to be put directly into a sharps box without resheathing them (see Figure 3-2). A more costly endeavor might be the risk associated with backing the fire apparatus into the bays. To avoid the risk, the department could design future stations with drive-through bays.

■ Note Any development of risk control measures must involve input from the employees who are affected.

Risk Control

If the risk cannot be avoided, then the next best effort that can be put forth is to control the risk or minimize the chance that it will occur. There are many ways to develop risk control measures and there may be several ways to control a particular risk. Any development of risk control measures must involve input from the employees who are affected. Risk control measures that are not or cannot be followed are useless.

■ Note The principle behind this thinking is that any interruption in one of the factors will prevent the injury or risk from occurring.

Interrupting the Accident Sequence One method of developing risk control is to view the injury process as a series of interruptible events. The accident sequence can be described in terms of five factors: the social environment, human factors, unsafe acts or conditions, accident, and the injury. The principle behind this thinking is that any interruption in one of the factors will prevent the injury or risk from occurring. Heinrich makes this theory analogous with dominos standing on end. If one domino is pushed, they all will fall. But if the center domino is removed, the falling will stop there.

In using this concept to generate risk control measures, the safety manager can look at past and predicted future risks and break the sequence down into events. Then these events can be classified according to the five factors in the sequence and control measures developed.

Let us take, for example, back injuries that occur from lifting heavy equipment from overhead compartments on the apparatus. The safety committee breaks down the events that lead up to the injury, then each event is classified. Table 3-2 shows the result of this process. This table is illustrative and not exhaustive.

Once the group has brainstormed all events and classified them, control measures can be developed. In this case, procedures could be written that require two persons to lift heavy equipment, apparatus can be designed for heavy equipment to be stored low, training could be undertaken to improve back strength and develop proper lifting techniques. Although this example is simple, the process can be used on much more complex problems.

Haddon matrix
a 4 × 3 matrix used to help analyze injuries and accidents in an attempt to determine processes designed to reduce them

matrix
a chart used to categorize actions or events for analysis

Note
Human factors are related to the employee or person involved in the injury. Human factors can contribute to the injury from three perspectives: poor attitude, inadequate training and education, or lack of fitness for the task.

The Haddon Matrix Another method of developing risk control strategies can be borrowed from a long-time injury prevention expert and developer of the **Haddon matrix,** Dr. William Haddon. Haddon based the **matrix** on the process of analyzing injuries. The matrix implies that there are four components of the injury process and three time intervals. The four components are the human factors, the energy vector, the physical environment, and the social environment. Human factors are related to the employee or person involved in the injury. Human factors can contribute to the injury from three perspectives: poor attitude, inadequate training and education, or lack of fitness for the task. The energy vector is what causes the injury. For a burn it would be the heat of fire or steam; for a back strain it might be the patient or piece of equipment, and for a laceration it could be the glass from a window or jagged metal on a damaged automobile. The physical environment is where the risk is likely to occur. This factor includes weather conditions, seating area on fire apparatus, a natural disaster, or placement of equip-

Table 3-2 *Classification of events leading to back injury.*

Event	Social Environment	Human Factor	Unsafe Act or Condition	Accident	Injury
Location of equipment on apparatus			X		
Fitness of the firefighter		X			
Not wanting to ask for help	X				
Muscle strain					X
Overexertion				X	

ment on the fire apparatus. The fourth component is the social environment, which involves the organization's climate, values and beliefs of employees, and peer pressure.

The time interval of preevent is studied to try to prevent the injury or risk from occurring in the first place. The event interval focuses on the risk at the time it is occurring; the strategies are aimed at reducing the exposure or minimizing the risk. Postevent is after the event has occurred and focuses on early notice that an event has occurred, proper care if necessary, and proper rehabilitation and recovery.

Risk control measures can be generated using this matrix. Although some of the control measures that could be generated may not be practical, many will be. A good way to use this tool is to select priority risk and have a group, often the safety committee, brainstorm control methods that would fit into each of the twelve blocks in the matrix. The committee or group can then go back and eliminate those that are not practical and further research those that are.

A result of this process is shown in Table 3-3, which depicts a partially completed matrix attempting to minimize burn injures after a brainstorming session by department officers and the safety program manager. This table is illustrative, not exhaustive. Local conditions and procedures may not allow for many of the measures to be implemented.

Table 3-3 *Example of a completed matrix.*

	Human	Energy	Physical Environment	Social Environment
Preevent	Train firefighters and company officers to recognize flashover conditions Have incident commanders make a risk analysis before committing to interior attacks on building not worth saving	Do not go in	Separate from cause Use proper hose stream and adequate gallons per minute Follow good tactics including ventilation	Education of public to be safer with fire Less peer pressure about wearing personal protective equipment (PPE)
Event	Train on survival techniques and rapid escape Have a rapid intervention team at the ready Wear a self-on PASS device	Wear full PPE	Have an escape route	Less aggressive attacks
Postevent	Training in burn care	Make others aware of fireground or interior conditions	EMS at scene for rapid care and transport	Proper follow-up care and rehabilitation

The Haddon thinking goes yet another step in the risk control strategy process. Once the measures that are practical are selected, the measures are further divided as to whether they are passive or active, and voluntary or mandatory. Passive here means that you do not have to do anything special for the control measure to occur, active means that you must physically do something. Voluntary measures allow people to choose, whereas mandatory measures are mandated by laws, rules, or regulations. It is evident that the most effective measures are those that are mandatory yet passive. Using PASS devices is common throughout the fire service. When they first came out, some individuals purchased them for themselves. This risk control method would have been considered voluntary and active. It is hard to predict how effective it was. Now PASS devices are a mandatory part of safety standards. However, until recently they required an active process, namely turning it on. Again, effectiveness was in question, because many were not activated while fighting a fire. Recent improvements have the PASS device turn on with the breathing apparatus, hence a passive process and an increase in effectiveness.

Some of the risk control measures from the burn example from Table 3-3 are further categorized in Table 3-4. This table is also for illustrative purposes and is not to be considered exhaustive. Note that some control measures may fit into more than one category.

Using the 4 × 3 and 2 × 2 matrix can be a valuable tool to assist the safety program manager in the development of risk control methods. It is easy to understand and provides the categorization necessary to formulate effective strategies.

Risk Transfer

risk transfer
the process of letting someone else assume the risk; for example, buying auto insurance transfers the consequences of an accident to the insurance company

Risk transfer in emergency services is most commonly associated with the purchase of insurance policies. The transfer of risk does nothing to prevent the risk and only limits the amount of financial exposure for the organization. The trans-

Table 3-4 *A completed matrix comparing voluntary and mandatory passive and active measures.*

	Voluntary	Mandatory
Passive	EMS at scene for rapid care and transport	Wear a self-on PASS device
Active	Use proper hose stream and adequate gallons per minute Have an escape route	Wear full PPE Use proper hose stream and adequate gallons per minute Wear full PPE

for of risk would do very little in terms of the time it takes to hire a replacement firefighter for one that was injured and retired on a disability pension. Risk transfer is closely related to the safety and health program, as the amount that is paid for insurance is based on previous experience, including injuries, accidents, and other risks.

Risk transfer through insurance is common. I do not know of any organizations, or individuals for that matter, that do not carry some type of insurance. Insurance carriers will be interested in the operation of the department from a risk management standpoint, not only to reduce claims, but in today's competitive environment many organizations bid their insurance coverages. The insurance company, therefore, also has an interest in keeping premiums low. Expect an insurance company to look favorably on things like driver training programs, standard operating procedures, drug-free workplace policies, and safety committees, to name a few.

However, the issue of risk transfer through insurance is a complex one, well beyond the scope of this book so only a brief explanation of different types of insurance is provided.

Workers' Compensation Insurance Workers' comp insurance is the insurance that covers employees injured while on the job. Each state has different requirements for workers' compensation so there is a great deal of variation in requirements and benefits. Generally, workers' compensation provides for the payment of medical expenses required after an illness or injury suffered on the job. Workers' compensation also pays a wage to the employee to cover time off following a job-related illness or injury. Since the wage that workers' compensation provides is only a portion of the employee's normal salary, there is incentive to get back to work. Some employment agreements provide for 100% of salary for a specified period of time after an injury or illness suffered on the job. In this case, the employer is required to make up the difference and may carry a secondary insurance policy.

Management Liability This insurance, sometimes referred to as errors and omissions, is designed to cover the actions, or lack thereof, of employees while performing duties on behalf of the employer.

General Liability This insurance is designed to protect the organization from a property loss. Although coverage varies, protection against theft, fire, storms, and such are usually provided.

Vehicle Insurance Vehicle insurance covers the organization's vehicles from damage or theft. Vehicle insurance policies also protect the insured from liability caused by the vehicle or its operator. Vehicle insurance generally does not cover an on-duty employee from an injury standpoint. Required medical treatment after a collision is covered by workers' compensation.

Regardless of an organization's safety record or safety and health plan, risk transfer is necessary. The cost of a single incident, such as a vehicle collision, could easily bankrupt a department without insurance. If injuries occurred, there would be property loss, medical expenses and lost time from injured employees, and liability for injuries to occupants of the other vehicles involved.

Summary

Risk management is a three-step process that requires risk identification, risk evaluation, and risk control. Risk identification, the first step, allows the safety manager to identify risks both from a historical basis and to predict those that are likely to occur in the future. Risk identification is done by analyzing local experience, identifying trends, conducting a safety audit, and reviewing the previous injury experience.

Risk evaluation is a measure of both frequency (how often a risk could occur) and severity (how bad it will be if it does). Together the evaluation of risk helps the safety manager set priorities for risk control programs

Risk control is the third step in the risk management process. Risk control can be accomplished by avoiding the risk, controlling the risk, or transferring the risk. In emergency services, risk avoidance has narrow application but in some cases can be useful. Risk control, however, can be used aggressively. The tools used to develop risk control strategies are interruption of the accident sequence and the Haddon matrix. Interruption of the accident sequence requires the classification of the accident events in order to develop measures to interrupt the process at some point. The Haddon matrix is a 4 × 3 matrix that considers four factors of three time intervals. The Haddon process further classifies the prevention measures into a 2 × 2 matrix as being active or passive and voluntary or mandatory.

Risk transfer is the process of transferring the risk to someone else. For the purpose of this text, the transfer would be to an insurance company. There are several types of insurance such as workers' compensation, vehicle damage and liability, property damage and loss, and errors and omission. Requirements for insurance are varied and beyond the scope of this text. Readers are encouraged to check local and state laws for insurance requirements.

Concluding Thought: Safety and health program management is not only subjective. It has objective components that are the foundation for local decision making.

Review Questions

1. List, in order, the three steps in the risk management process.
2. What four areas can be examined when performing the risk identification process?
3. Explain the difference in the terms *frequency* and *severity*.
4. Give an example of risk avoidance.

5. List and explain, in general terms, the four kinds of insurance coverage.
6. What are the twelve components of the Haddon matrix?
7. Which of the following would be most effective as a risk control measure?
   A. Voluntary/Active
   B. Mandatory/Passive
   C. Mandatory/Active
   D. Voluntary/Passive
8. What are the four classifications in the accident sequence, according to the Haddon matrix?
9. Give two examples of human factors in the Haddon matrix.
10. Define the social environment.

Activities

1. Select an injury class from your department's statistics, for example, back injuries while advancing hose lines. Apply the Haddon matrix and develop some risk control measures.
2. Using the same risk as in Activity 1, apply the interrupting-the-accident sequence process. Compare the similarities and differences. Which process produced better results?

Preincident Safety

Learning Objectives

Upon completion of this chapter, you should be able to:

- Describe safety considerations in the emergency response station.
- Explain safety considerations as they apply to the emergency response vehicle.
- List the components of an effective response safety plan.
- Describe the components of a preincident planning process.
- List the information that should be provided by the preincident plan.
- Describe the considerations for safety while training.
- Define the components of a wellness/fitness plan.
- Describe the considerations for interagency coordination as it applies to health and safety.

CASE REVIEW

In August 1996, the International Association of Fire Chiefs (IAFC) Fire-Rescue International was being held in Kansas City, Missouri. On Saturday, August 24, the annual Combat Challenge was scheduled. On that day, members of the International Association of Fire Fighters (IAFF) Local 42 and other IAFF locals protested the event with informational pickets. The picketing was intended to allow firefighters to exercise their right to express grievances, primarily about the way the Combat Challenge allegedly has been used in a noncompetitive, job threatening manner in some departments. The picket line grew to nearly 200 supporters. Following reports of intimidation by picketers, assaults, and damage to competitors' vehicles, the IAFC canceled the competition.

After this incident, IAFC President Chief R. David Paulison was quoted as saying "We must now take the opportunity for management and labor to explore the issues and try to reach an agreeable approach to firefighter physical fitness." And that is exactly what happened. Fire chiefs and union leaders from ten fire departments from across the United States and Canada were selected to sit on this committee and develop a comprehensive physical fitness program.

One year later, the IAFF/IAFC joint physical fitness committee presented a program at both the IAFC's Fire Rescue International and at the IAFF's Redmond Symposium. The program is a complete nonpunitive program that includes requirements for medical examinations, physical and mental fitness, rehabilitation, and wellness.

The complete program and training package are available from the IAFC or through IAFF locals.

INTRODUCTION

During the risk identification process, a number of risks can be identified that occur before an incident does. These risks focus on issues relating to the workplace, the equipment or vehicle being used, and the human resource that is using it or working in the workplace. In this chapter, the focus is on risks that occur prior to an incident or at the time of an incident. Many of these risks are the easiest to manage, because, with few exceptions, the risk occurs under a controlled situation or environment.

Preincident health and safety can be divided into seven categories:

1. Station safety
2. Apparatus safety
3. Response safety
4. Safety while training
5. Fitness/wellness
6. Interagency cooperation in safety related matters
7. Preincident planning

STATION SAFETY

Consideration for safety at the station is grouped into two broad categories: design and ongoing operations. The design categories deal with fire station design prior to construction or at the time of renovation (see Figure 4-1). Ongoing operations deals with safety and health issues from a day-to-day operational basis, such as wiping up wet spots to prevent falls.

Design

The fire or emergency medical services (EMS) station should be designed with safety as an important consideration. First, the designer must ensure that the station is built to comply with all applicable standards and codes. The National Fire Protection Association (NFPA) standard 1500 has a section regarding fire stations and the station should also comply with local building codes and the local fire prevention code (Figure 4-2). Stations should be designed with private fire detection and protection systems such as sprinkler systems, smoke and heat detectors, and carbon monoxide detectors, if a source is present.

Once the design is in compliance with local codes and standards, it is important to look at the physical design. Proper routes of travel must be provided from living areas to the apparatus bays. These routes of travel must be direct, without obstruction, and lighted for night operations. Sleeping areas must be designed so that, after dark, low-level lighting is provided to prevent tripping and fall injuries. If the station is multistory and chutes or poles are used, there must be protection from employees falling down them by accident, and new employees must be trained on the proper use of these devices. The bays should be provided with proper drainage to limit standing water, and walk areas should be

Figure 4-1 *There are many different designs for fire stations. Each should consider responder safety during day-to-day operations.*

Figure 4-2 *Fire stations should be designed to meet all local building and fire codes.*

painted using a nonslip paint. Bathroom floor and shower areas should be nonslip surfaces as well.

■ **Note**
By OSHA regulation, there must be an area for the cleanup of materials contaminated with bloodborne pathogens.

Storage areas for protective clothing must allow for ventilation. Unless contracted out, a laundry facility may be provided. By Occupational Safety and Health Administration (OSHA) regulation, there must be an area for the cleanup of materials contaminated with bloodborne pathogens.

The bays should have some form of exhaust emission system. Exhaust systems may be the type that attach to the vehicle's exhaust by a tube and are vented to the outside. A second system vents the entire apparatus bay area, using roof or wall-mounted exhaust fans and louvers in the bay doors. This system is useful in colder climates where small equipment, such as rescue tools and saws, must be run inside the bays. Yet another exhaust elimination system is not part of the station, but attaches and remains on the emergency vehicle exhaust.

■ **Note**
The station design process should include input from the safety committee and the crews themselves.

The station design process should include input from the safety committee and the crews themselves. The crew will be able to pinpoint safety issues in the design of a building.

It clearly is much easier and less costly to design the previously mentioned features into a new building prior to construction. However, most of the facilities that the safety manager deals with today have long been constructed. Some are more than 100 years old. The safety manager, in conjunction with fire department management and the safety committee, must prioritize improvements and try to retrofit the station with the necessary safety features over several budget years. Modifications that are required by regulation may have to be made early on.

Sufficient parking should also be provided, particularly for volunteer departments where members respond to the station for the apparatus. Responders should be able to park where they can exit their vehicles and access the station safely.

Ongoing Operations

Of additional concern regarding station safety, and probably more important, are the procedures and safety measures relating to the day-to-day operations of the station and the inhabitants, the responders. The station is generally a controlled environment, so injuries that occur at the station could often easily have been prevented.

■ Note

The fire station should have a complete fire inspection at least annually in accordance with local code.

First and foremost, the fire station should have a complete fire inspection at least annually in accordance with local code. Fire protection systems should be certified as recommended or required in the code. Smoke and carbon monoxide detectors should be checked according to manufacturer's recommendation. Generally, the on-duty crew can perform these inspections, sometimes with assistance from the fire prevention division and using standard forms (see Figure 4-3). Local OSHA or state departments of labor will usually help a department identify safety problems in the station, as will many insurance carriers. One insurance company even provides inspection forms.

Aside from the fire safety concern, station inspections that look for hazards and safety issues should be conducted daily. Part of the shift change process should require the on-coming supervisor to walk the station area and look for safety or health problems. These might include wet spots on floor, the proper operation of an emergency bay door stop, or burnt out lights in a bedroom. Further department procedures should forbid unsafe acts such as creating tripping hazards with debris or electrical cords. The procedures can also be proactive, such as using Wet Floor signs after mopping a floor. It is interesting that on the fireground a firefighter will climb 15 feet to a second floor window for an operation and using a leg lock, lock into the ladder. But in the station that same firefighter will climb 25 feet on a stepladder to change a light bulb and not consider tying off. Procedures and training should prohibit this type of behavior. Some examples of station safety policies are provided in Text Boxes 4-1 and 4-2.

TEXT BOX 4-1 EXAMPLE POLICY.

FACILITY SAFETY

Purpose

Fire stations, like all structures, are subject to fire and safety hazards. This standard is designed to keep fire stations as safe as possible to protect the lives and health of the personnel within.

Definition

Facility safety is the result of regular inspections for various hazards and conscientious personnel who are always on the lookout for dangerous situations and acts.

Objectives

- To schedule regular monthly and annual inspections of all fire stations.
- To provide for the immediate correction, or to be corrected as soon as possible, of all hazards and code violations found during said inspections.
- For all personnel to become safety conscious about their individual stations.

Station Inspection

Annual inspections. An annual inspection of each station will be done by a company in-service inspector (minimum) and will be accompanied by the shift captain or his designee. The inspection will be done on the first Thursday of September using the standard Safety Survey Form. Copies of the form will be given to the Fire Prevention Bureau and to the shift captain or his designee. Corrections will be made as soon as possible.

Monthly Inspections. A monthly walk-through inspection will be done on the first Thursday of each month using forms provided. The intent of this inspection is to identify unsafe conditions, check that exit lights and smoke detectors work properly, and that extinguishers and the Ansul System are properly charged. Department fire sprinkler forms shall also be completed and attached to inspection form. Hazards or violations will be reported in writing to the Fire Prevention Bureau and the shift captain or his designee. Corrections will be made as soon as possible. The shift captain or his designee will accompany the inspector on these inspections.

TEXT BOX 4-2 (COURTESY FLORIDA DEPARTMENT OF LABOR).

Office Safety (All Personnel)

A. Do not connect multiple electrical devices to a single outlet.

B. Do not use extension or other power cords that are cut, frayed, or damaged.

C. Close file and desk drawers when unattended. Do not open more than one drawer at a time and close it when done.

D. Put heavy files in bottom drawers of file cabinets to prevent cabinets from tipping over.
E. Do not tilt your chair back on two legs.
F. Do not use chairs, boxes, or improvised climbing devices.
G. Turn off the machine and disconnect electrical power before attempting to adjust or clear electrical office equipment.
H. Do not remove, bypass, or tamper with electrical equipment fuses, switches, or safeguards.
I. Do not place your fingers in or near the feed of a paper shredder. Verify guards are in place and working prior to use.

Station Safety (All Personnel)

A. Mop/clean up oil, hydraulic fluid, water or grease from apparatus floors and accesses, immediately, upon detection.
B. Safety goggles or other provided eye protection shall be worn when operating power equipment.
C. Do not run electrical and other cords across doorways, aisles, between desks, or create trip hazards.
D. Pick up all foreign objects, such as pencils, bolts, and similar objects, from floors to prevent slipping.
E. Clean up all spills immediately, especially wet spots around drink and coffee machines, in bathrooms, kitchen, and hallways.
F. Use stepladders only on a firm level base, opened to the full position with spreaders locked.
G. Do not leave tools in work areas, walkways, or stairways.
H. Do not walk across wet or oily areas. Avoid the hazard by walking around it.
I. Avoid walking across areas that have been freshly mopped.
J. Do not run on stairs or steps or take two at a time.
K. Use handrails.
L. Do not block your own view by carrying large objects.
M. Do not jump from trucks, platforms, scaffolds, ladders, roofs, or other elevated places, regardless of height.
N. Do not take shortcuts. Use aisles, walkways, or sidewalks.
O. Do not jump from, to, or between elevated areas.

STATION INSPECTION RECORD

STATION ____________ SHIFT ____________ DATE ____________

CAPTAIN ____________ BATTALION CHIEF ____________ DIVISION CHIEF ____________

✓ - O.K
X - SEE REMARKS

1. PERSONNEL

A. Uniform
B. Protective Clothing
C. Grooming & Cleanliness
D. Driver's License
E. I.D. Card
F. E.C. Card
____
____

5. SAFETY

A. Overhead Storage
B. Caution Signs
C. Safety Bulletin Board
D. Tool Storage
E. Combustibles Storage
F. Insecticide Chemical Stor.
G. SCBA Tests
H. Extinguisher (s)
I. Smoke Detector (s)
J. RADEF Equipment

6. EXTERIOR

A. Paint
B. Doors
C. Sidewalks
D. Ramps
E. Parking Area
F. Yard
G. Hose Tower
H. Gas Pump
I. Flag Pole

2. OFFICE

A. Files
B. Maps
C. Log Book
D. Shift Change Log
E. Desk
F. Bulletin Board
G. Tourist Information
____

3. ENGINE HOUSE

A. Floor
B. Work Bench
C. Hose Rack
D. Storage
E. Turnouts
F. Apparatus
____

7. INTERIOR

	Office	Kitchen	Day Room	Dormitory	Rest Room	Heater Room		
A. General Appearance								
B. Cleanliness								
C. Floors								
D. Windows								
E. Lights								
F. Furniture								
G. Storage Area (s)								
H. Woodwork								
I. Walls								
J. Ceiling								
____________								
____________								
____________								
____________								

4. PROGRAMS

A. Hazard Reduction
B. In Service Inspections
C. Pre Fire Plans
D. Water Systems
____

REMARKS ____________

Fire 580 2415 016 Catalog # 9020 (Rev. 2/90)

Figure 4-3 *An example of a station inspection record.*

APPARATUS SAFETY

Safety involving apparatus also has some preincident considerations. Like stations, apparatus considerations can also be categorized into design and ongoing operations. Design concerns are those that are analyzed and controlled at the time of purchase. Ongoing concerns are those that affect day-to-day operations. One major issue associated with apparatus is driving and responding to emergencies; that issue is addressed in the next section.

Design

Emergency service vehicles should be designed to meet applicable standards. There are NFPA standards that govern requirements for fire apparatus (see Chapter 2 for listing), many of which are based on firefighter safety needs. There are also requirements placed on the vehicles by the Department of Transportation, because the vehicle, in most cases, will travel on public roads. Ambulances also have federal regulations from the Department of Transportation (see Figure 4-4).

Although designing and meeting the minimum requirements stated in the

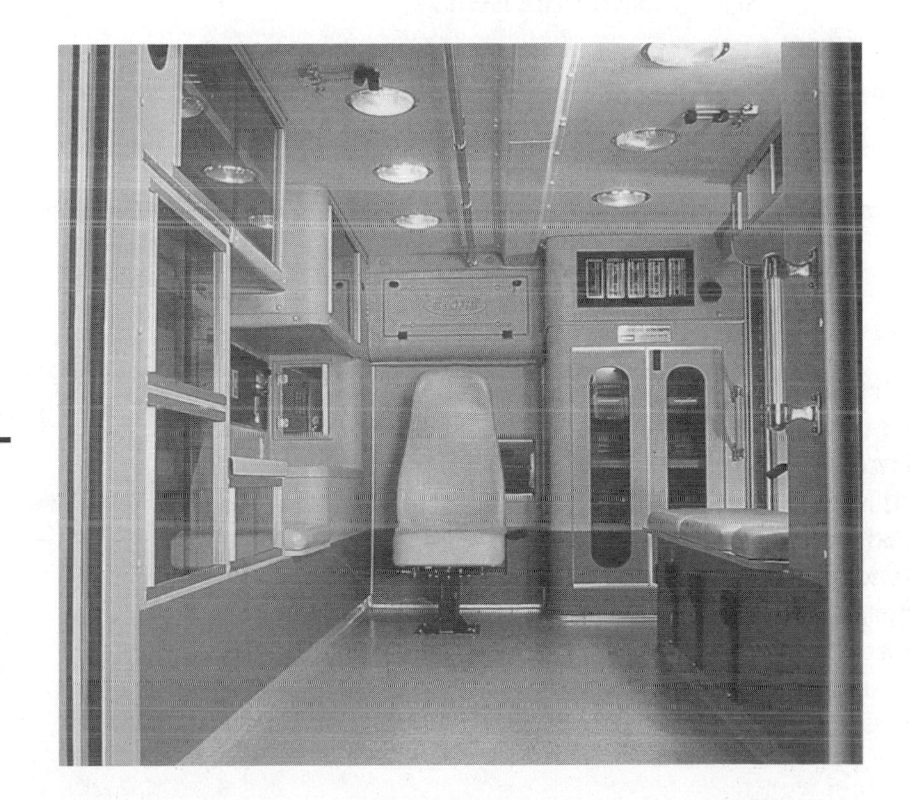

Figure 4-4
Ambulance design is governed by the department of transportation in the KKK specifications. Here is an example of interior layout. (Courtesy of E-One, Inc., Ocala, FL.)

standards provides some level of safety, the safety manager and safety committee should have direct involvement in the design process. A risk identification and analysis should be part of the design phase. For example the placement of equipment affects compartmentalization requirements and location. How will ladders be stored? If attached to the side of the apparatus, will they be too high to be safely removed? How about hose bed height for both supply line and preconnects? Can they be safely deployed and, just as important, safely repacked? The standard allows for the first step into the cab to be 24 inches from the ground. Is this low enough for a department that has identified ankle injuries caused by getting out of the cab as a risk? Or could another step be installed which might lower the step level to 18 inches? What about the design of the patient compartment of the ambulance? Again, the standards provide some level of safety, but the unit should be designed around local needs and conditions.

Remember, manufacturers of emergency vehicles are willing to do just about anything possible to design a vehicle to meet the needs of their customer. However, many times modifications cost money and may be deleted in favor of other requirements for the apparatus. The safety manager's responsibility is to perform a thorough risk analysis and support each and every safety consideration.

Ongoing Operational Concerns

preventive maintenance (PM) program
ongoing program for maintenance on vehicles, designed to provide routine care, oil changes and the like, as well as catch minor problems before they become major

Most departments keep emergency vehicles for a number of years. A lot can be said for a vehicle that is designed properly, but most of the safety issues relating to apparatus are of the day-to-day operational nature.

Many of the ongoing operational safety objectives can be accomplished by having a strong **preventive maintenance (PM) program** (see Figure 4-5). The PM program should include daily checks of emergency lights, fluids, and the operation of vehicle components including brakes, pumps, transmission, and aerial ladders. The daily check should be performed by the apparatus operator assigned that day. The operator should be trained to look for maintenance problems before they become worse or cause an accident. Using the apparatus operator also allows the operator to become familiar with the vehicle and vehicle components.

■ Note

Lack of preventive maintenance reduces the useful life of a vehicle, increases maintenance costs, and can have a significant negative effect on the safety of those relying on the vehicle.

A more thorough check should be performed weekly, including running small tools, cleaning the apparatus thoroughly, checking tire wear and pressures, and back-flushing pumps. A monthly check may go even further, requiring the performance of such tasks as changing water in the water tanks.

Another component of the PM program is a semiannual trip to the maintenance professional. This trip can provide for scheduled fluid changes and visual checks of brake linings and other critical vehicle systems. In the case of fire apparatus, pumps and ladders must be tested annually as well.

This type of preventive maintenance program can be applied to any size or any type of department. Whether the apparatus is a fire truck or an ambulance, the same concepts apply. Lack of preventive maintenance reduces the useful life

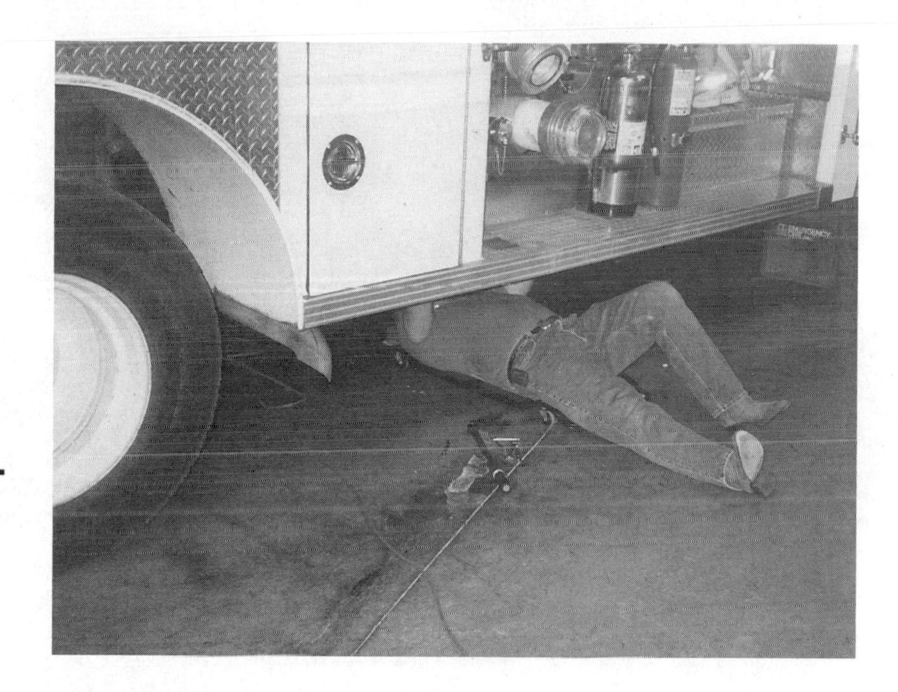

Figure 4-5 *A preventive maintenance program is integral to having a safe response vehicle.*

of a vehicle, increases maintenance costs, and can have a significant negative effect on the safety of those relying on the vehicle.

Standard operating procedures and training can also help in the day-to-day operational safety process. Procedures identifying the minimum number of personnel to perform certain tasks have been successful in reducing injuries. For example, for reloading the hose, the procedure may require four persons to be on hand: two in the hose bed, one on the back step, and one on the ground. Keeping the apparatus running board free from debris or fluids prevents slip and fall injuries. Proper lighting when working around the vehicle can also reduce injuries.

Apparatus safety procedures should also address hearing conservation. The United States Fire Administration (USFA) has published a manual to assist emergency service organizations in developing a hearing conservation program (see Figure 4-6).

National Association of Emergency Vehicle Technicians (NAEVT) organization that bestows professional certification in many areas for persons involved in emergency service vehicle maintenance

Apparatus safety must include risk identification and evaluation, and the implementation of control methods that can only be designed at the local level. Text Box 4-3 outlines some apparatus safety considerations, although some are related to response. Another resource is the **National Association of Emergency Vehicle Technicians** (NAEVT), which can provide valuable information on apparatus maintenance and safety operating practices. The address is listed in the further resource section of Appendix II.

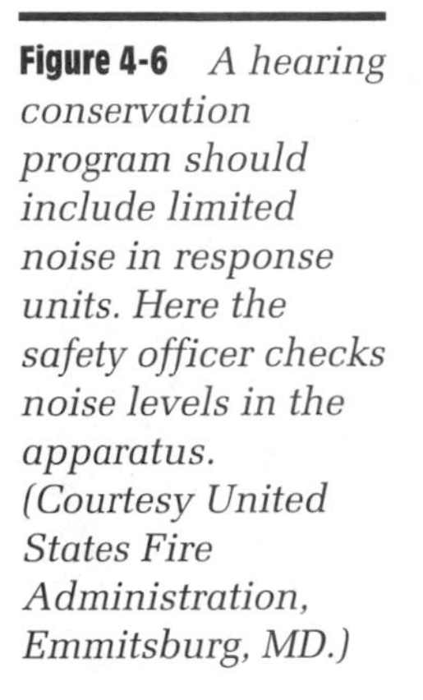

Figure 4-6 *A hearing conservation program should include limited noise in response units. Here the safety officer checks noise levels in the apparatus. (Courtesy United States Fire Administration, Emmitsburg, MD.)*

TEXT BOX 4-3 EXAMPLE POLICY.
APPARATUS SAFETY (COURTESY OF PALM HARBOR FIRE RESCUE)

355.0 Purpose

The purpose of these standards is to establish safety procedure in regard to apparatus practices thereby reducing the potential of accidental injury or death to fire department personnel and the general public.

355.1 Vehicle and Equipment Inspection Procedures

A. In order to ensure vehicle readiness and presence of essential equipment, all apparatus and equipment shall be regularly inspected and/or tested by personnel at intervals prescribed on the existing Vehicle Inspection Form. Daily vehicle and equipment checks shall be performed as close to 0800 hours as practicable and their timely completion is to be superseded only by emergency calls or by the order of the officer in charge.

B. Any apparatus or equipment deficiencies shall be noted on the departmental Vehicle/Equipment Deficiency Form. The officer in charge shall also be verbally notified of any such deficiencies. The officer in charge shall then take necessary action to affect repairs or remedy the deficiency. The equipment or apparatus shall be placed out of service should the deficiency be deemed serious or hazardous.

C. All vehicles and equipment shall be outside of the bays while operating (weather permitting).

355.2 Driver's Responsibility

A. It is the responsibility of the driver of each fire department vehicle to drive safely and prudently at all times. Vehicles shall be operated in compliance with traffic laws pursuant to F.S. 316. Emergency response does not absolve the driver of any responsibility to drive with due caution. The driver of the emergency vehicle is responsible for its safe operation at all times. The officer in charge of the vehicle has responsibility for the safety of all operations.

B. It is the driver's responsibility to ensure that all passengers are safely aboard the apparatus and seated prior to moving the vehicle.

C. When responding to an emergency, emergency warning lights must be on and sirens must be sounded to warn drivers of other vehicles.

D. Intersections present the greatest potential danger to emergency vehicles. When approaching and crossing an intersection with the right-of-way, drivers shall not exceed the posted speed limit.

When approaching a negative right-of-way (red light, stop sign, or yield), the vehicle shall come to a complete stop. The vehicle may proceed only when the driver can account for all oncoming traffic and all lanes yielding right-of-way.

E. In order to avoid unnecessary emergency response, the following rules shall apply:

1. When the first unit reports on the scene with "nothing showing," "investigating," or the equivalent, any additional units shall continue emergency, but shall not exceed the posted speed limit and shall come to a complete stop before proceeding at every negative right-of-way intersection.
2. The first arriving unit will advise additional units to respond non-emergency whenever appropriate.

F. During an emergency response, fire vehicles should avoid passing one another. If passing is necessary, arrangements should be conducted through radio communications.

G. The unique hazards of driving on or adjacent to the fireground require the driver to use extreme caution and to be alert and prepared to react to the unexpected.

H. When stopped at the scene of an incident, vehicles should be placed to protect personnel who may be working in the street and warning lights shall be used to make approaching traffic aware of the incident. At night, vehicle-mounted floodlights and any other lighting available shall be used to illuminate the scene.

I. Backing. When the backing of apparatus is unavoidable, the driver shall use a guide to back and shall not proceed to back unless the guide is clearly visible in the mirror(s) being used for visualization. Illumination of the backup area with floodlights and/or flashlights shall be used in times of darkness or limited visibility. If no guide is available, the driver shall dismount and walk completely around the apparatus before proceeding to back. Audible backup alarms shall be used during any backing procedure.

355.3 Seating/Seatbelts

A. Personnel are to ride in a seated position at all times that the apparatus is in motion and only in seats or positions where seatbelts are provided. Seatbelts are to be utilized at all times when vehicles are in motion. Under no circumstances will personnel be permitted to ride in a standing position or on any sideboards or tailboards of any apparatus.

B. EXCEPTIONS to 355.3(A) above:

1. Seatbelt use will not be required during actual fire combat from exterior positions on the brush or water trucks during brush fires. For this reason, brush/water truck drivers shall exercise extreme caution with regard to speed, terrain, overhead clearance, etc. If terrain is extremely rough or other conditions exist that would render exterior positioning unsafe, the rider shall dismount from the vehicle and either proceed on foot or take a seated position in the interior of the cab. This exception does not exempt personnel from the required use of seatbelts in brush/water apparatus during routine or emergency over-the-road operations or while traveling point-to-point during off-road operations.
2. Fire department personnel riding in ground or air ambulances, or any other vehicles shall wear seatbelts when possible.
3. It is obvious that seatbelts are impractical during repacking of hose in the hose bed of apparatus. This should not be attempted without adequate personnel.

C. When responding to calls requiring the donning of protective clothing, donning shall be completed prior to response since this procedure would be difficult to complete while seatbelts are engaged.
D. If a call requiring the donning of protective clothing is received during routine driving of apparatus, the driver shall proceed to a safe location off the roadway where personnel may don protective clothing prior to continuing response. It is not recommended that personnel wait until arrival at the scene to don protective clothing.
E. Donning of air packs while apparatus is in motion shall be done only in apparatus provided with seating which permits donning while seat belts are engaged and proper seating/posture can be maintained. In apparatus where this is not possible, personnel shall don air packs upon arrival at the scene.
F. Safety bars (e.g., Man Savers) shall be in the down position at all times and shall not be disabled or altered in such a manner that would defeat intended function.
G. A common cause of ankle/knee injuries is forward (front-facing) dismount from apparatus. All personnel, especially when wearing air packs or carrying heavy loads, should back down off of apparatus using hand-hold handles to assist.

RESPONSE SAFETY

Although closely related to apparatus safety in terms of the vehicle being safe, response safety has a broad range of applications to emergency service health and safety and therefore is included under a separate heading. Response safety involves everything from the selection and training of the drivers and operators to the physical environment in which the response takes place. For the purpose of this section we assume that the station and the vehicles have already undergone the risk management process so neither is considered under this heading.

One response safety concern for volunteer and combination departments is responders driving to the station or to incidents in their private vehicles. Strict policies must be in effect for these situations and the drivers must be aware that often they are not in a marked vehicle and therefore the public may not be aware of them as responders.

The first consideration in response safety is the selection, training, and capabilities of the driver-operator. Apparatus driver selection should consider the human aspects such as attitude, knowledge, mental fitness, judgment, physical fitness, and past driving record. Prior to driver selection, many departments perform some type of testing that measures some or all of these characteristics. Further, prior to or after selection, the driver should be trained in emergency vehicle operations in a vehicle similar to the one he or she will drive and operate. This training program should address legal issues, physical forces including controlling the vehicle on a variety of road surfaces, safe stopping distances, vehicle maintenance, departmental safety procedures, general safe driving practices, and specific safe driving practices based on the geographic area and environment. This training must have both classroom and practical components. A driving course is a common approach to the practical phase of this training, but provisions must be made for different driving surfaces, including wet roads. A commonly used driving obstacle course setup is shown in Figure 4-7.

■ Note

Apparatus driver selection should consider the human aspects such as attitude, knowledge, mental fitness, judgment, physical fitness, and past driving record.

The department's insurance company may also offer driver training resources; some even have complete programs including an instructor. The emergency vehicle driver must have the capabilities to control the vehicle and react

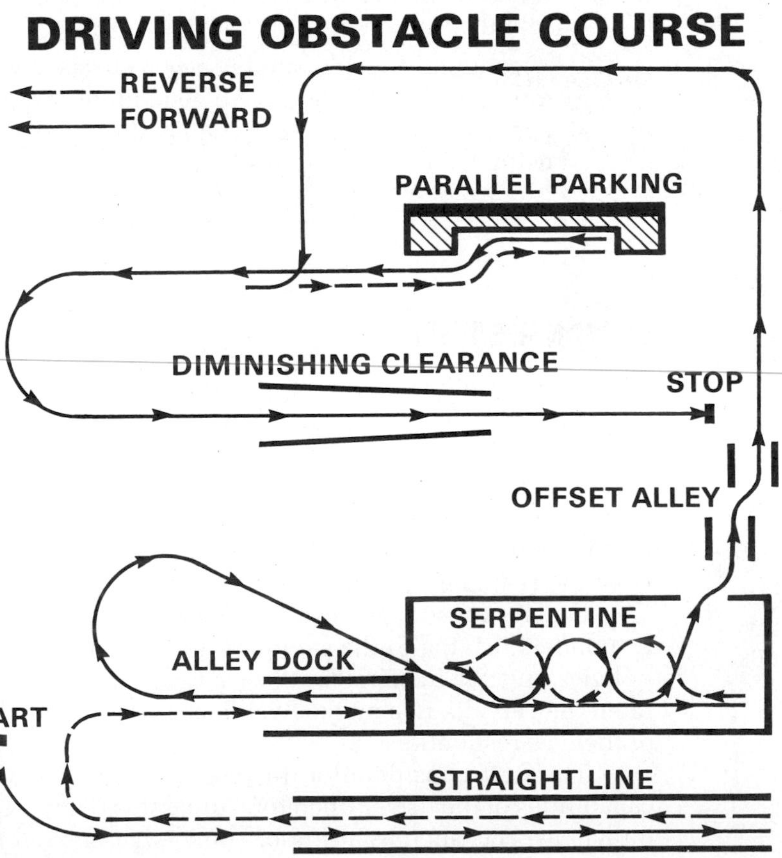

Figure 4-7 *A typical obstacle course for emergency services driver training. (Courtesy USFA.)*

quickly and appropriately to changing situations. Excessive speed, reckless driving, and failing to slow down or stop going through intersections have been the causes of many emergency vehicle accidents. Policies that prohibit operating an emergency vehicle while under the influence of drugs or alcohol, including prescription medications, must also be in place and enforced.

Once the driver is selected and trained, the program must focus on the response. Consideration for procedures that require the use of seats and seatbelts is necessary. The habit of riding tailboards or sidesteps is unacceptable and for the most part has stopped. However, standing in the jump seat area still occurs. The supervisor on the emergency vehicle should be required to ensure that all occupants are properly seated and belted prior to the response.

The type of response also has safety implications. An emergency response using warning devices is more dangerous than a response in which the vehicle flows normally with traffic. The safety manager, in conjunction with the department management, should review the types of incidents that the department responds to and determine which may not require emergency response. The St. Louis fire department (SLFD) has taken this approach and reduced the number of intersection accidents greatly. Since implementation of this program, the SLFD has only had one serious accident in three years or 180,000 alarms. Text Box 4-4 describes this program.

TEXT BOX 4-4 ST. LOUIS FIRE DEPARTMENT NONEMERGENCY RESPONSE POLICY.

Companies and chief officers will respond on the quiet, no lights or sirens, to the following incidents:

- Automatic alarms
- Sprinkler alarms
- Natural gas leaks
- Wires down
- Calls for manpower
- Flush jobs
- Lockouts
- Investigation drums, barrels, unknown odors, and such
- Carbon monoxide detector alarms
- Rubbish, weeds, and dumpster fires
- Move-ups to city or county fire stations

- Broken sprinkler or water pipes
- Smoke detectors
- Manual pull stations
- Refrigerator doors
- Plugging details
- Assist the police
- Keys in running vehicles

If a response is dispatched on the quiet and additional information is received by fire alarm indicating that life is in danger, persons are injured, there is a working fire, and so forth, fire alarms will upgrade the response to "urgent," lights and sirens.

■ **Note**

The environment in which the response takes place can also be analyzed and risks identified.

The environment in which the response takes place can also be analyzed and risks identified. Traffic patterns and common response routes should be studied. Apparatus accidents have occurred between two fire department vehicles responding to the same incident from different locations. This risk can be reduced with better radio communication and preincident procedures for response. Target intersections can also be identified. A problem intersection where emergency vehicle collisions or near misses have occurred should be avoided. However, this choice is seldom reasonable. The department could look at intersection control. One product on the market causes the light to turn green for the responders after activation by the vehicle's siren. This system costs about $6,000 to install, a lot less than the price of one accident, let alone more than one.

Procedures may require that the vehicle come to a complete stop at every intersection or not exceed the speed limit in residential neighborhoods. The environment is dynamic, traffic patterns change with time of day, weather can change quickly. Road construction should be planned for and response routes evaluated. Areas along the response route that have poor visibility should be identified prior to the response. Railroad crossings, particularly those that are not controlled should be noted on maps, or the drivers should be trained in their locations. In 1989, a Catlett, Virginia, fire truck was struck by a train at a grade crossing. Two of the responders were killed and three injured. The report on the collision recommended that the department adopt procedures relating to response that requires crossing of railroad tracks.

PREINCIDENT PLANNING

The value of preincident planning cannot be underscored enough. Not only from a safety standpoint, but also from an operational standpoint. Having preincident

■ Note

In case of a fire, preincident planning allows the fire department to fight the fire before it happens.

boiling liquid expanding vapor explosions (BLEVES)
occur when heat is applied to a liquified gas container and the gas expands at a rapid rate while the container is weakened by the heat; when the container fails, a BLEVE is said to occur

plans available is like a coach having the playbook at a game. In the case of a fire, preincident planning allows the fire department to fight the fire before it happens. From a safety standpoint preincident planning can be used to identify and evaluate potential risk long before an incident occurs. For example, preincident planning may reveal that an occupancy uses and stores combustible gases. The gases are compressed and stored in a storage room. Having identified this hazard, responders can adjust tactics to limit their exposure to potential explosions or **boiling liquid expanding vapor explosions** (BLEVES), should a fire occur.

To be effective the preincident planning process should contain the following components:

- The preincident plan should be on a form used departmentwide.
- Preincident planning should be done by the responders so that they can become familiar with the building during the preparation of the plan.
- The process should provide for updating the plan at given time intervals. A good goal would be an annual update. Remember, occupancies change and renovations occur.
- Target hazards should be identified and receive priority in the planning process. Target hazards may include:
  - Health care facilities
  - Large industrial occupancies
  - Facilities using or storing hazardous materials
  - High-rise buildings
  - Malls or other high occupancy locations, such as office complexes or stadiums
  - Schools
  - Hotels, motels, and apartment buildings
- The preplan should include text as a reference, a site plan, and a floor plan.
- The preplan should show or provide the following information:
  - Location
  - Construction features
  - Building emergency contacts
  - Occupancy type and load
  - Hazardous materials or processes
  - Location of utilities
  - Location of entrances and exits
  - Ventilation locations
  - Nearest water supply
  - Need for specialized extinguishing agents

- Private fire protection equipment and systems on site, including fire department connections
- Mutual aid plans
- In addition, some preplans provide for placement of apparatus on a first alarm assignment.

Although preincident planning components address many of the needs of responders to fires or hazardous material emergencies, preplanning also applies to EMS. For example, EMS responders may want to preplan alternate accesses to a nursing home after hours. Are there alternate means of egress? Will a stretcher fit in the building's elevator? What about access to locked apartments for a medical emergency when a person cannot make it to the door. Many times complexes have on-site maintenance or security with keys, which is easier and safer than breaking down the door and does not upset the owner as much.

Preincident planning is an excellent tool when considering preincident safety and risk identification. However, preincident planning is commonly limited to target hazards and many of the profession's injuries and deaths occur in single family dwellings.

SAFETY IN TRAINING

Based on the 1996 NFPA injury report, 6,200 firefighter injuries occurred during training. This number represents 7.1% of the total injuries. Training evolutions are generally created to simulate actual events, but the training evolutions must be controlled so that injuries can be further prevented. The NFPA standard 1403, *Live Fire Training Evolutions in Structures,* was created as a result of a number of injuries and deaths that occurred during live fire training. The standard followed closely the deaths of three Milford, Michigan, firefighters who were killed during a training burn at an acquired structure. The firefighters were engaged in a training exercise in a two-story structure and were trapped on the second floor after a fire was started on the first floor. They did not have adequate personal protective equipment (PPE), nor the safety of a handline. Training evolutions must be designed and conducted only after a thorough risk analysis including identification and control of the risks. Injuries and deaths associated with training are avoidable, and the safety manager should adopt a zero tolerance level in this regard.

Note
Injuries and deaths associated with training are avoidable, and the safety manager should adopt a zero tolerance level in this regard.

The NFPA standard 1403 sets forth procedures and requirements for live fire training, some of which are also applicable to nonfire training evolutions. The 1403 standard addresses the following subjects:

- Assignment of safety officers
- Preburn walk-through of the structure
- Building acquisition procedures
- Securing utilities

- Environmental impact of the burn
- Building construction and conditions
- Use of proper protective equipment
- Proper water supply including backup lines and supply
- Use of an incident management system
- Instructor-to-student ratio
- On-scene emergency medical care

Many of the requirements addressed in this standard can become part of a department's standard procedures for training evolutions. For example, how often is it permitted that hose streams are operated with just helmets or helmets and gloves, or worse yet, no protective clothing? In this case, the training is not simulating actual conditions and the risk of injury from the hose is much greater. Is it department policy to assign a safety officer during a hazardous materials exercise, or during a multicompany drill at the drill tower using a ladder? It should be policy. Do you provide rehabilitation and EMS at the site of outside evolutions? Do you follow the same safety procedures during EMS training as you would in the field in terms of disposal of blood products and needles? Do you require protective equipment during extrication exercises? Repeating cutting up cars in the summer in full PPE causes trainees to want to cut corners. If a safety officer is watching and prohibits such action, it may save an injury. The procedures should also provide for an accountability system at multicompany training evolutions similar to that used at real incidents. This not only allows for practicing the system but affords another level of safety for the participants.

■ Note

The procedures should also provide for an accountability system at multicompany training evolutions similar to that used at real incidents.

EMPLOYEE WELLNESS AND FITNESS

An employee wellness and fitness program is an essential component of the safety and health plan. Fitness and wellness programs can be developed locally using in-house talent or by contracting with professional exercise and wellness consultants. A number of resources are available to assist departments in developing these programs. The IAFF/IAFC has a joint initiative, the USFA publishes a fitness coordinator's manual, local universities often have an industry hygienist who may be able to help, and department physicians may be able to provide information on program development. A number of components are necessary for a program to be successful and effective.

It is difficult to separate the terms *wellness* and *fitness*. Generally the occupational safety and health programs each have components that can fit in these categories, but for the program to be effective there must be tight integration. Wellness is a broad term that would include components relating to medical fitness, physical fitness, and emotional fitness. Wellness programs should also include behavior modification programs including weight reduction, tobacco use cessation, nutrition, and stress reduction.

A successful wellness-fitness program must also have provisions for after-injury rehabilitation and care. A rehabilitation program should be designed to ensure that the employee has a path to rehabilitation and return to duty with the employee's safety in mind. Rehabilitation programs should include input from the department doctor and physical therapy agencies familiar with the job of an emergency responder. **Alternative duty programs** are also helpful to permit a safe return. These programs allow an employee to work during his or her recovery, but within restrictions based on physician's orders. Periodic evaluations of the employee's condition is also an integral part of the program.

alternative duty programs
sometimes called light duty or modified duty, these programs allow an injured employee to return to work, but with restrictions for some period of time while recuperating

Medical Fitness

Medical fitness can be assessed by performing yearly medical exams. Although recommendations differ as to how often these exams should be conducted for different age groups, it is agreed that annual medical exams are a necessity. The NFPA standard 1563 addresses medical requirements for the fire service and recommends intervals based on age. Members assigned to specialized teams, such as hazardous materials response, are required, by regulation, to have annual exams, regardless of age. The medical exam should be conducted by a physician who is very knowledgeable in occupational issues and specifically the work and hazards faced by emergency responders. Text Box 4-5 provides the tests recommended by the IAFF/IAFC wellness-fitness initiative.

TEXT BOX 4-5 MEDICAL TESTS RECOMMENDED BY THE IAFF/IAFC WELLNESS-FITNESS INITIATIVE.

- Medical History Questionnaire
- Vital Signs
- Hands-On Physical
  - Head, eye, ears, nose, and throat
  - Neck exam
  - Cardiovascular—submaximal tests
  - Pulmonary—including spirometry and chest X ray (repeat X rays recommended every three years)
  - Gastrointestinal
  - Genitourinary
  - Rectal—including digital rectal exam
  - Lymph nodes

Musculoskeletal
Body composition
Laboratory Tests

White blood cell count	Potassium
Platelet count	Albumin
Glucose	Heavy metal screening
Sodium	Red blood cell count
Total protein	Triglycerides
Cholesterol tests	Creatine
Differential	Carbon dioxide
Liver function tests	Calcium
Blood urea nitrogen	Urinalysis

Vision Test
Hearing Test
Cancer Screening

Breast exam	Fecal occult blood testing
Prostate specific antigen	Pap smear
Testicular exam	Skin exam
Mammogram	

Immunizations and Infectious Disease Screening

Vaccinations	Varicella
Hepatitis C	Hepatitis B virus
HIV	Polio
Tuberculosis	

aerobic fitness
a measurement of the body's ability to perform and utilize oxygen

cardiovascular fitness
fitness levels associated with the cardiovascular system, including the heart and circulatory system

body composition
a measure to show the percentage of fat in the body; there are certain published parameters for what is considered average or normal

Physical Fitness

Speaking generally, physical fitness requirements for emergency responders should address three areas: **Aerobic** or **cardiovascular fitness,** muscular fitness, and **body composition.** These components interrelate and are required of the emergency responder. A person can have a great deal of muscular fitness and be very strong, but have a low level of cardiovascular fitness or endurance. Clearly the work of an emergency responder requires both.

Cardiovascular fitness and body composition can be assessed as part of the medical exam. Muscular fitness can be assessed using a battery of tests. The handgrip dynamometer can be used to measure grip strength, the arm-flex dynamometer measures the maximum force generated from arm flexion, and the leg-extension dynamometer measures the maximum force that can be generated from extension of the legs. Push-ups and sit-ups can also be used to measure muscular strength and endurance.

Once the fitness level is identified, a program should be developed by the department fitness coordinator on an individual basis. For example some employees may lack cardiovascular endurance, while others may have excellent cardiovascular endurance but lack muscular fitness. The program should also provide for on-duty time and the facilities for members to work out and maintain the required level of fitness.

Emotional Fitness

Emotional fitness is also a necessary component of the program. A responder who is under undue stress, from whatever cause, will not be able to perform effectively and is prone to injury. Emotional fitness can be improved by having programs in place for employee assistance. These programs should be accessible to employees and their families on a confidential basis. Often employers pay the cost for a fixed number of visits through a contract with local behavior care providers. Family members are encouraged to participate too, as often the emotional problems are related to family and home life. Services provided by some employee assistance providers include:

Note
Employee assistance programs may also be part of the discipline process in an organization.

- Drug or alcohol abuse treatment
- Tobacco use cessation
- Family problems
- Financial problems
- Stress management
- Critical incident stress management

Employee assistance programs may also be part of the discipline process in an organization. If an employee's work performance takes a sudden downturn, the reason may be an emotional problem. Requiring mandatory participation in an employee assistance program can be used as a step in the discipline process and contribute to the future of the employee and the department.

critical incident stress management (CISM)
a process for managing the short- and long-term effects of critical incident stress reactions

Clearly a factor in the emotional wellness of an emergency responder is **critical incident stress management** (CISM). Having a CISM program in place contributes to the health and safety program. This subject is presented in Chapter 8 with postincident safety issues.

INTERAGENCY CONSIDERATIONS

Another preincident safety and health issue that is sometimes overlooked is the relationship between the responder's agency and other agencies that we may deal with. Foremost should be the relationship with mutual or automatic aid providers. In most emergency response agencies, some mutual or automatic aid agreement exists. The safety program manager must know how the other departments operate. Are they safety conscious, do they use the same incident management system, the same accountability system? Do their operations comply with recognized standards? Are the communications systems compatible? Responders also work closely with law enforcement agencies. Law enforcement agencies commonly respond to many incidents with the fire or EMS responder at vehicle crashes, incidents requiring crowd or traffic control, incidents resulting from criminal activities, and in some jurisdictions as first responders to all medical incidents. In some areas the law enforcement agencies may provide rescue or aeromedical evacuation. This response creates the need for integration in response to some incidents and a preincident understanding of roles and relationships (see Figure 4-8).

unified command
used in the IMS when two or more jurisdictions or agencies share incident command responsibilities but do so in conjunction with each other

Having an understanding of each agencies priorities and roles prior to an incident occurring is critical to safety and health activities and must be preplanned. In some situations there will not be clear lines of authority and responsibility. For example, in the 1985 MOVE incident in Philadelphia, sixty-three homes were burned. Both police and fire were held responsible. In such situations, the incident management system must provide for **unified command,** which, simply put, provides for an incident commander from both law enforcement and the emergency response agencies. These commanders work in unity to bring the incident to a safe conclusion.

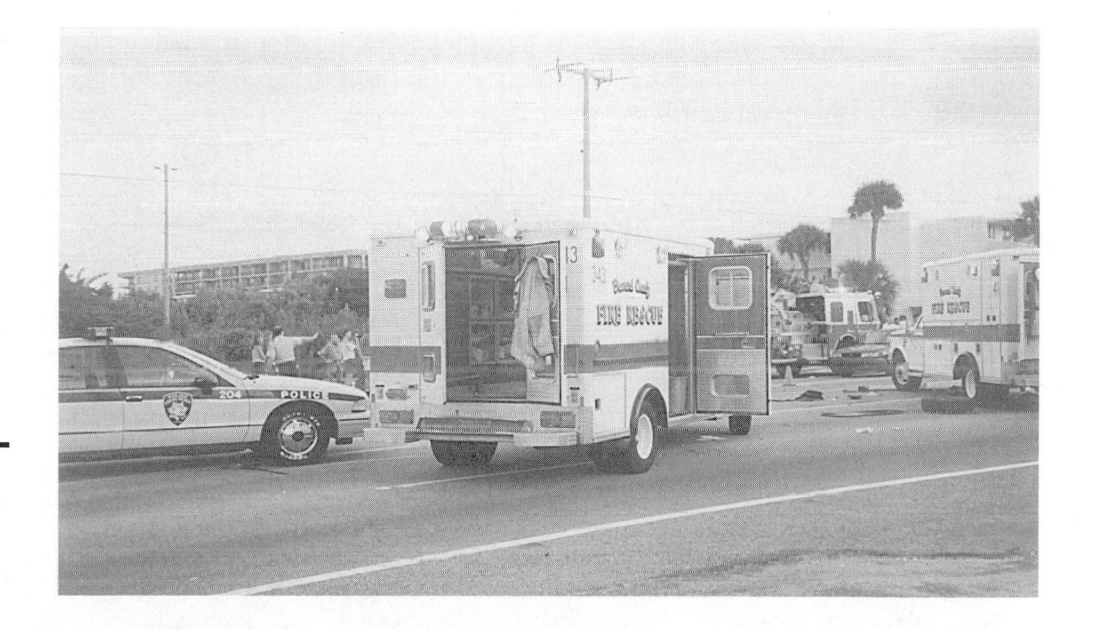

Figure 4-8 *At this vehicle crash, three different public safety agencies are working together.*

Other issues can be preplanned, for example, do the emergency responders wait for police clearance before entering a scene where violence, shootings, or stabbings are possible? Do the law enforcement agencies provide traffic control at an interstate accident, or is their priority to reopen the highway? What is the expectation of the emergency responder in terms of protection by law enforcement during periods of civil unrest?

This relationship is a two-way street. Law enforcement may ask the emergency responders to avoid using red lights and sirens when nearing a scene of a shooting in the street in order not to draw a crowd, making the officers' jobs unsafe. Or the aeromedical helicopter operated by the police may need some specific ground support for landing zones. All of these considerations can be dealt with on a local level and in many different ways. The important point is to realize that the issues must be decided upon prior to the incident and with the safety of ALL responders in mind.

■ Note

The important point is to realize that the issues must be decided upon prior to the incident and with the safety of ALL responders in mind.

Although the emergency responders will probably interact most with mutual aid and law enforcement in terms of interagency cooperation, other agencies are involved as well. Hospitals provide initial treatment or medical surveillance after exposures to infectious diseases. The relationship with the local emergency room should be an integral part of the safety and health programs. Again what are the expectations of the emergency responder agencies and the expectation of the hospital? Depending on the department situation, a communication center may be operated by a separate agency where safety and health concerns for responders can be preplanned. The local health department can be a resource for information on infectious diseases.

Depending on local conditions, safety and health for the emergency responder can be impacted, either positively or negatively, by a preincident relationship with outside agencies. This relationship can go far not only from a safety and health standpoint, but also from the standpoint of effective and integrated operations.

Summary

Preincident safety encompasses a number of components including station safety, apparatus safety, response safety, preincident planning, safety during training, wellness/fitness programs, and interagency relations. Station and apparatus safety can be subdivided into issues relating to design and ongoing, day-to-day operations. Response safety includes driver selection, driver training programs, and policies affecting response. Preincident planning is a tool that can be used by responders to effect safer operations. Building characteristics, occupancy, and particularly hazards can be identified prior to an incident. Means of access and tactical considerations can also be discussed prior to an emergency and under better conditions.

Safety during training is a concern in the overall health and safety program. Training is often a simulation of real conditions and is a controlled, although somewhat unpredictable, environment. All training evolutions should be conducted with participant safety as the number one priority.

An integral part of the safety and health program is a program to ensure the highest level of fitness and wellness for the responders. Wellness is a broad term that encompasses a number of components, including medical, physical, and emotional fitness. Medical fitness can be assessed with medical exams. Physical fitness assessments are necessary and individual programs are developed for responder improvement. An employee assistance program is vital in ensuring emotional fitness.

Emergency responders work alongside and rely upon a number of other agencies. Strong interagency coordination and preincident defined roles and expectation can be helpful to prevent injuries and death. Interagency coordination also leads to a more effective operation.

Concluding Thought: Safety is a state of mind. Many components can be prepared for before an incident.

Review Questions

1. Which of the following would not normally be considered as part of the daily vehicle apparatus checks?
   A. Back flushing the pump
   B. Checking the emergency lights
   C. Checking the oil level
   D. Checking the seat belts
2. List the areas of concern when selecting emergency vehicle driver/operators.
3. Preventive maintenance programs should require biannual trips to the maintenance professional.
   A. True
   B. False

4. What is one option that exists for traffic light control at intersections?
5. List five of the lab tests that should be included in an annual medical examination.
6. Muscular fitness is one of three types of fitness important to the emergency responder.
   A. True
   B. False
7. List two external agencies that a department may want to coordinate with before an incident.
8. For the purpose of preincident planning which of the following would not be considered a target hazard?
   A. A hospital
   B. A lumber yard
   C. A railyard
   D. A two-unit duplex
9. List three components of an effective vehicle preventive maintenance program.
10. List three methods of determining muscular fitness.

Activities

1. Review your department's policy on medical examinations. Compare the policy to the standard and the recommendations in this text. How does your exam compare? If you do not have a medical examination program, research programs and write a draft policy.
2. Review your department's standard operating procedures for response to incidents. Is the policy current? Could the emergency response mode be reduced to some incidents? Is there a policy or procedure for the selection and training of drivers?
3. Review your department's standard procedures for safety in training. Does your live fire training comply with NFPA 1403? With application from 1403, could improvements be made to day-to-day training activities?
4. Identify your department's relationship with other agencies. Are there standard procedures or other prearranged agreements? Are the policies in effect or do they need to be established?

Safety at the Fire Emergency

Learning Objectives

Upon completion of this chapter, you should be able to:

- List the three incident priorities.
- Explain the relationship between the three incident priorities and the relationship to health and safety.
- Discuss in general terms the hazards faced by responders to fire incidents.
- List the components of personal protective equipment used for fire incidents.
- Discuss the need for and the components of an effective accountability system.
- Discuss the types of and the relationship between incident management systems and health and safety of the responder.
- Define the need for and uses of a rapid intervention team.
- Discuss the components of incident rehabilitation.

CASE REVIEW

On Valentine's Day, 1995, at 12:22 A.M., the Pittsburgh Fire Department responded to a house fire in the 8300 block of Bricelyn Street. Upon arrival, the firefighters found a fire in a three-story wood frame structure with a basement. The building was constructed on a grade, which gave the appearance from the front that the house was only two stories. The initial tactic was to mount an interior attack and the first arriving engine stretched a line to the interior to search for the fire and to effect extinguishment.

During this operation, a stairwell used by the firefighters collapsed. After the collapse, firefighters were reported to be trapped inside. Because several companies had by then arrived and were working in the interior, supervisors were unable to quickly assess which companies were missing.

Shortly after the report of firefighters being trapped, several firefighters were rescued through an exterior window in the rear. At this time there was no report accounting for all firefighters on the scene, and, as a result, the incident commander was not aware that three other firefighters remained trapped in the building. The three remaining firefighters were discovered after most of the fire was knocked down and smoke ventilated, about 1 hour after the firefighters entered the building.

After the National Fire Protection Association (NFPA) investigation was performed, several factors were identified as contributing to this tragedy. Two of the factors cited were lack of procedures that allow for a quick accounting of firefighters from the beginning of the incident and the use of rapid intervention teams for unexpected situations that occur on the fire scene, enhancing the chance of rescue and survival.

INTRODUCTION

As seen in Chapter 1, most firefighter deaths and injuries occur while operating at fire scenes as a result of different causes, ranging from stress to building collapse. The safety and health program manager must have a good knowledge of the hazards faced at fires and procedures that will ensure safer operations.

Forming an understanding of the incident priorities and their relationship with operations and hazards will improve the safety manager's ability to design a training program, purchase equipment, and implement procedures targeted toward injury reduction.

INCIDENT PRIORITIES AND SAFETY

Priorities for incident management can be categorized and placed in descending order of seriousness into life safety, incident stabilization, and property conservation. These priorities are the first thing that the incident commander must consider in the incident, and they should be continuously reevaluated as the incident progresses.

Figure 5-1 *Safety officer should perform a size-up based on incident priorities.*

Life Safety

■ Note
Life safety is the group of activities that ensures that the threat of injury or death to civilians and emergency personnel is reduced to the absolute minimum.

Life safety is the first incident priority as it is based on the premise that life safety must be maximized. Life safety is the group of activities that ensures that the threat of injury or death to civilians and emergency personnel is reduced to the absolute minimum. This is done by limiting the exposure of danger to the absolute minimum. In a simple situation, it may mean that emergency responders wear protective clothing while operating. In a complex incident involving a fire in several buildings, it may require the evacuation of several blocks of residences and setting up a collapse zone. Regardless of the situation or how the priority is met, life safety is always number one throughout the incident.

Incident Stabilization

■ Note
Incident stabilization is the group of activities required to stop additional damage or danger.

The second priority is incident stabilization. During this time, attempts are made to solve the problem. Incident stabilization is the group of activities required to stop additional damage or danger. Stabilization could mean a quick interior attack on a room and contents fire or a defense attack on a fire to stop the spread from building to building. Size-up helps to determine initial strategy (see Figures 5-1, 5-2).

■ Note
This group of activities is commonly called salvage or stopping the loss.

Property Conservation

Property conservation is the final incident priority and involves reducing the loss to property and the long-term health and welfare of the people affected. This group of activities is commonly called salvage or stopping the loss (see Figure 5-3).

Figure 5-2 *Fire officer performs size-up upon arrival at the incident scene.*

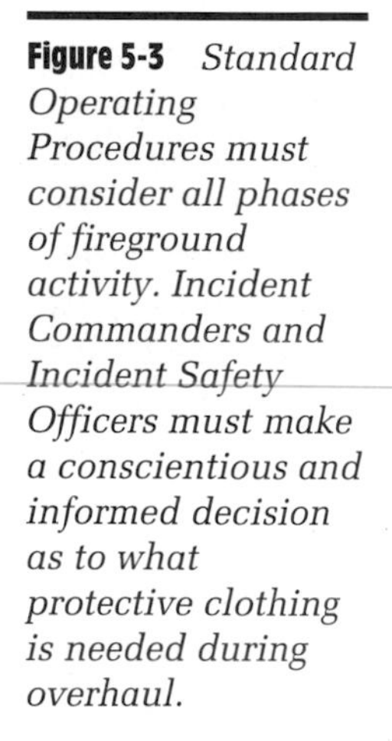

Figure 5-3 *Standard Operating Procedures must consider all phases of fireground activity. Incident Commanders and Incident Safety Officers must make a conscientious and informed decision as to what protective clothing is needed during overhaul.*

In a simple incident, the use of a salvage cover may be all that is necessary. In a large incident, the emergency response agency may be in property conservation mode for hours or days. Property conservation, or stopping the loss, is rapidly becoming the fire service's benchmark for excellence.

Relationship between Incident Priorities and Safety

Remember that life safety is always the number one priority and all operations must be developed based on this priority. As an incident priority, life safety applies to both civilians and emergency responders.

Recall from Chapter 3 that risk management began with risk identification. This first step, identification, should be the basis for risk management on the emergency scene. Identification is the same as size-up or risk assessment. Because the life safety of the responders is the number one priority, incident operations and strategies can be developed that provide the highest level of safety to responders. In other words, for incident stabilization to occur, the incident commander should have considered life safety to the responders in formulating the strategy.

Risk assessment of the emergency scene can be based on the following philosophy:

We will risk a lot to save a lot,
We will risk little to save a little,
We will risk nothing to save nothing.

In more useful terms, the incident commander would have few qualms about sending a search and rescue crew into a building fire if there were a possibility that people were inside. However, arriving at the same fire, meeting the occupants outside, and confirming that they are all out may produce a different, safer strategy. Although some argue that a building is never unoccupied until a search proves it, more prudent and safety-minded incident commanders are changing strategies to make them more compatible with the risk.

HAZARDS FACED BY RESPONDERS

In order to do an effective risk assessment and to implement an effective strategy, it is necessary to understand the risks present at fires. For the purpose of this chapter, fire incidents are divided into four groups:

1. Structure fires
2. Transportation fires
3. Outside structure fires
4. Wildfires

Fires related to specialized incidents such as hazardous materials are discussed in Chapter 7.

Structure Fires

Structure fires account for the majority of fires that injure and kill most firefighters. Depending on the makeup of a local community, these may or may not be the largest number of fires responded to. For discussion purposes, structure fires can be further subdivided into one- and two-family dwelling, multifamily dwellings, commercial, and industrial in terms of occupancy. Table 5-1 compares hazards related to each of these occupancies. A similar table could be completed after a risk identification process in your jurisdiction. The hazards listed are general enough to apply to buildings of different size, configuration, and height. However, a number of hazards are common to any structure fire. Some hazards are

Table 5-1 *Hazards associated with various types of occupancies.*

Hazard	Residential 1 and 2 Family	Multistory Residential	Commercial	Industrial
Life hazard	High	High	Medium, depends on time of day	Medium, depends on time of day
Structure failure	X	X	X	X
Large area operations		X	X	X
Exposures threatened by fire		X	X	X
Back draft	X	X	X	X
Flashover	X	X	X	X
Explosions			X	X
Chemical	X	X	X	X
Improper storage	X	X	X	X
Long operations/need for rehabilitation		X	X	X

■ Note
Risk assessment should consider the construction type and the anticipated reaction of the construction methods and materials to fire and heat.

back drafts
occur as a result of burning in an oxygen-starved atmosphere; when air is introduced, the superheated gases ignite with explosive force

flashover
a sudden, full involvement in flame of materials and gases within a room

mushrooming
heat and gases accumulate at the ceiling or top floor of a multistory building then back down; can be prevented with vertical ventilation

rollover
the rolling of flame under the ceiling as a fire progresses to the flashover stage

directly related to the structure in terms of construction. Risk assessment should consider the construction type and the anticipated reaction of the construction methods and materials to fire and heat. For example, it is no secret that by design, lightweight, wood truss roof assemblies have failed early in operations, and the incident commander should consider this in the risk analysis. Hazards relating to the structure that should be considered are:

- Types of loads and transmission of loads (fixed, live, dead, etc.)
- Construction type
- Building material types
- Structural elements including beams, girders, columns, connectors, and support systems
- Avenues for fire and smoke spread

Again, risk identification is necessary for local risk management.

Because many structure fires are extinguished from the inside of the building, fire behavior characteristics must also be considered. The incident command and safety manager must be familiar with the signs and potential outcome of situations such as **back drafts, flashovers, mushrooming, rollover,** and **heat**

Figure 5-4 *Proper tactics, including ventilation, reduce the chance of back drafts.*

heat transfer
the transfer of heat through conduction, convection, radiation, and direct flame contact

transfer (see Figure 5-4). Through training, education, and experience, the incident commander can make risk assessments and minimize the exposure of responders to these potentially deadly hazards.

Transportation Fires

Fire involving transportation includes vehicle fires, both auto and truck, train, aircraft, and ships. Each of these forms of transportation presents unique problems to the emergency responders (see Figure 5-5). With the exception of auto and truck each form of transportation fire could make a textbook by itself. However,

Figure 5-5
Transportation fires can range from the family car to large trucks, railcars, airplanes, or ships.

like structure fires, some generalizations can be made and local risk identification can assist in assessment of the local risk.

Generally, hazards associated vehicle fires involve the vehicle itself, the fuel used to power the vehicle, and what cargo the vehicle is carrying. Fire involving the vehicle components themselves may weaken structural components and allow them to fail. When they fail under heat or fire, some vehicle components may explode or become flying missiles. For example, bumpers on vehicles that are designed to withstand low speed collisions without damage work by using a piston. When exposed to heat, the gases in the piston expand and ultimately the piston can fail, sending the bumper hurling toward responders. The fuel that powers a vehicle is a flammable or combustible liquid, which adds to the fuel of the fire. And although usually only thought of in the case of trucks, the cargo is also of concern in any vehicle accident. How many people take their small barbeque tank to the refill station in the trunk of their car? Or how many travel home from the pool store with pool chemicals? Vehicle fires large or small require the same risk assessment and application of incident priorities as do structure fires. On-scene risk assessment is necessary to ensure application of appropriate strategies.

■ Note
Vehicle fires large or small require the same risk assessment and application of incident priorities as do structure fires.

Outside Structure Fires

Fires occurring outside of structures can range from dump fires to grass fires. Each can present different levels of hazards and risks. Dumpster or garbage fires require special attention, as the responder normally has little information regarding the contents of the containers. Some clues can present themselves, however. A dumpster behind a hardware store can lead one to assume that paints and other chemicals might be involved. Other possible contacts in dumpster fires include hazardous materials and bloodborne pathogens.

■ Note
Dumpster or garbage fires require special attention, as the responder normally has little information regarding the contents of the containers.

Outside fires also expose responders to certain topographic and environmental hazards, including weather and access hazards. These fires often do not have a civilian life hazard associated with them, so a good risk assessment in these situations is risk a little to save a little. This philosophy can go a long way in preventing injuries.

Wildfires

Some of the most publicized fires are wildfires. Whether the fire be in a forest in Wyoming or in the urban interface of California, these fire are large, long in duration, and produce enormous property and environmental loss. They also are responsible for a number of firefighter injuries and deaths each year. Some metropolitan areas never see a wildfire, while for some departments, wildland fires are very common. Because of the nature of these fires, commonly over large geographic areas and for long periods of time, strong incident management systems with good accountability procedures must be in place (see Figure 5-6).

Figure 5-6 *Wildfire in progress. Note the onlookers—another safety concern for responders.*

The crews fighting these fires are often very far from help, so it is of critical importance that the incident management system provide for long-term planning and logistical needs, such as food and rehabilitation of crews. Safety program managers and incident commanders who function in wildfire-prevalent areas must have knowledge of weather and the weather's effect on the fire. Included in weather are the effects of temperature, air masses, temperature inversions, relative humidity, wind velocity direction, and fuel moisture. For further risk assessment, consideration must be given to topography including elevation and slope, fuel including ground fuels, aerial fuels, structural fuels, and fire behavior. These incidents take the incident management system to high levels, often filling positions that were only designed for special wildland situations. Text Box 5-1 discusses safety tips for wildland firefighting.

TEXT BOX 5-1 SAFETY TIPS FOR WILDLAND FIRES.

Ten Fire Fighting Orders

1. Keep informed on fire weather conditions and forecasts.
2. Know what the fire is doing at all times.
3. Base all actions on the current and expected behavior of the fire.
4. Plan escape routes for everyone, and make them known.
5. Post a lookout where there is possible danger.
6. Be alert, keep calm, think clearly, act decisively.

7. Maintain communication with your crew, your supervisor, and adjoining forces.
8. Give clear instructions, and be sure they are understood.
9. Maintain control of your crew at all times.
10. Fight fire aggressively, but provide for safety first.

PERSONAL PROTECTIVE EQUIPMENT

Personal protective equipment (PPE) designed for fire fighting is one strategy in providing for minimizing exposure at the fire scene and addressing the priority of life safety as applied to responders. PPE for fighting structure and vehicle fires is essentially the same. Wildland fire fighting has a different PPE design, although many departments do not provide specialized PPE. Specialized PPE for incidents such as technical rescue and hazards materials is discussed in Chapter 7.

A discussion on PPE for response to fires can be broken down into three subject areas:

- Design and purchasing
- Use
- Care and maintenance

Standards from the NFPA (1900 series) set forth requirements for the purchase of fire fighting PPE, including minimum requirements for particular items of PPE. See Chapter 2 for specific standards and respective components of the PPE ensemble. The NFPA standard 1500 requires that emergency responders be provided with PPE based on the hazards of the particular work environment. Once provided, guidelines or procedures must be adopted governing the use of the PPE. Procedures must be in place governing a care and maintenance program, including inspections.

Design and Purchase

The NFPA 1500 standard requires that new fire-fighting PPE meet the current editions of the respective standards. Older gear must have met the standard in effect when purchased. Therefore, the purchasing agent for the department should reference the particular NFPA standard, in its entirety, in the specification for the PPE. The specification writing process, from a safety and health standpoint, should include input from the safety committee, or at a minimum, the safety program manager.

The commonly expected components of PPE for firefighting (see Figure 5-7) include:

Figure 5-7
A firefighter in full NFPA-compliant PPE.

- Approved fire helmet with protective eye shield
- Flame-resistant hood
- Turnout coat
- Turnout pants
- Firefighting gloves
- Firefighting boots
- Personal alert safety system (PASS) devices
- Self-contained breathing apparatus (SCBA)[1]

Use

Standard operating procedures (SOPs) should also be in place defining the use of the PPE. Although this seems quite obvious, it is helpful from the safety standpoint to require the use of full PPE under specific conditions. Examples might

[1] Although the SCBA unit is usually placed on the apparatus, many departments supply personal face pieces.

include that full PPE be used for interior structure fire fighting. But when is it permissible to remove the SCBA? During overhaul? After the fire is extinguished? The SCBA policy should require the SCBA mask be on until the safety officer or incident commander deems the atmosphere in the work area to be safe. This decision can only be safely made by using air monitoring equipment. Procedures can also define different levels of PPE for different types of fires. Is SCBA required for wildland fires? How about a dumpster fire? A firefighter would never approach a hazardous materials incident without wearing SCBA, yet would fight a dumpster fire behind a hardware store without SCBA. The use of PPE can be linked to common sense. However, like other components of a complete safety program, the proper use of PPE should be outlined in policies and procedures.

Care and Maintenance of PPE

■ Note
The safety program must have a component of PPE inspection procedures, repair procedures, and care.

The NFPA standards on PPE also provide guidelines for the care and maintenance of the equipment. The safety program must have a component of PPE inspection procedures, repair procedures, and care. Generally, the standards require following the manufacturer's recommendations.

Because the PPE is issued to individual members, often the care is also delegated to them. However, because this equipment is specialized, the department has an obligation to provide a process or the facilities for cleaning and minor repairs. In some departments, both of these functions are handled in house, while other departments contract with outside vendors to provide this service. If the department also has emergency medical services (EMS) responsibilities, the bloodborne pathogens procedures must also be considered in the cleaning of the PPE.

Although the care is delegated to the user, the safety program must include a procedure for periodic inspection to ensure the PPE is in good condition. Monthly inspections are recommended (see Figure 5-8). The monthly inspection can be performed by a station officer or, in some departments, the shift com-

Figure 5-8 *Company officer doing monthly inspection of PPE.*

mander. However, the safety program manager should also be involved in the process. A quarterly or semiannual inspection by the safety program manager or staff is a good idea. After each use, the PPE should be inspected by the user. Of course, SCBA and PASS devices should be checked and inspected at the beginning of each shift and after use. Figure 5-9 provides an example of a PPE inspection form.

EXAMPLE
PERSONAL PROTECTIVE EQUIPMENT RECORD

Employee Name: ____________________ Year: __________

Item	Inv #	Issued	July	Aug.	Sept.	Oct.	Nov.	Dec.
Helmet								
Hood								
SCBA Mask								
Mask Bag								
Safety Glasses								
Bunker Coat								
Gloves								
Bunker Pants								
Suspenders								
Bunker Boots								
Work Boots								
HEPA Mask								
			Inp/Res	Inp/Res	Inp/Res	Inp/Res	Inp/Res	Inp/Res

Inp=Inspection Date and Inspector Initials/ Res=Result of Inspection (Pass or Fail)

REMARKS

July: ____________________

August: ____________________

September: ____________________

October: ____________________

November: ____________________

December: ____________________

Figure 5-9 *A typical inspection check sheet.*

Table 5-2 Sample PPE record.

Assigned to: ID#:

Station #: Shift:

Activity	Helmet	Coat	Pant	Gloves	Hood	Boots	PASS	SCBA	Remarks (Use Back as Needed)
6 Month Clean 1/1/98	X	X	X	X	X	X			
Inspection 1/5/98	X	X	X	X	X	X	X	X	Pass battery due Hole left pant knee
Knee repaired 1/15/98			X						
Battery replace 1/15/98							X		
Annual Flow Test								X	Completed by Smith's SCBA shop

Records should be kept on all PPE. Included in the record should be assignment, dates when cleaned, date and type of repairs, and inspection dates. This record should be filed in the workplace of the person who has the PPE assigned them. Table 5-2 is an example of a PPE record.

incident management system (IMS)
an expandable management system for dealing with a myriad of incidents to provide the highest level of accountability and effectiveness; limits span of control and provides a framework of breaking the big job down into manageable tasks

INCIDENT MANAGEMENT SYSTEMS

A key requirement to maximize safety of the fire scene is to use an **incident management system (IMS).** A number of recognized systems are in use throughout the country. For the purpose of this text, it is not important *which* system is used, but that a system is used. Local conditions and procedures will dictate the type of system, but the system, to be effective, must have the following design characteristics:

- Must provide for different kinds of operations including, single jurisdiction/single agency response, single jurisdiction/multiagency response, and multijurisdiction/multiagency response.

- Organizational structure must be adaptable to any emergency.
- Must be usable and understood by all agencies in a particular geographic area.
- Must be expandable in a logical manner from the beginning of an incident.
- Must use common terminology.
- Must have integrated communications.
- Must maintain a manageable span of control.
- Must be used on all incidents!

Because this is not a text on IMS, the focus is on the role of safety within the IMS structure. The IMS structure provides for a command staff and general staff. Included in the command staff are three functions or positions: information, safety, and liaison. The general staff has four positions: operations, finance, logistics, and planning (see Figure 5-10). Each of these functions is performed at each and every incident. At a single unit response incident, the company officer may fill all the command staff roles; at a one-alarm house fire the shift commander may fill all the roles except for safety and information. At a multialarm commercial fire, administrative officers may fill in many of the command staff or general staff functions. At a several-week-long fire fight at a wildfire, all IMS positions may be filled and the system may be several layers deep.

■ Note To be effective, the safety officer must have expertise in fireground risk management.

From a safety and health perspective, the incident safety officer is part of the command staff. To be effective, the safety officer must have expertise in fireground risk management. The safety officer must also have authority within the IMS and direct communication with the incident commander. Requirements and roles of the incident safety officer are further described in Chapter 9.

■ Note At an incident scene, the safety officer or member of the safety sector or group must be given certain authorities.

At an incident scene, the safety officer or member of the safety sector or group must be given certain authorities. These include stopping, altering, and suspending operations that are determined to be unsafe. The safety officer must alert the incident commander of the situation and what action the safety officer

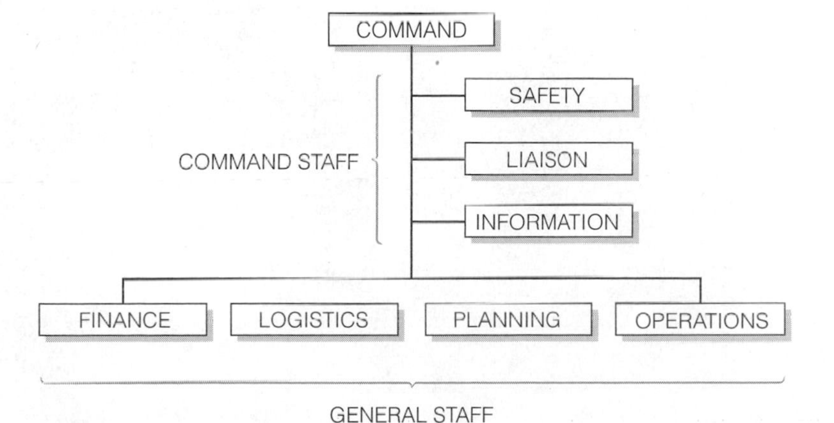

Figure 5-10 *Incident management system—command staff and general staff positions.*

freelancing occurs when responders do not follow the incident plan at a scene and do what they want on their own; or a failure to stay with assigned group

has taken. The safety officer working at a fire scene should be alert to a number of safety-related actions, proper PPE being worn, structural conditions, establishing collapse danger zones, overseeing the accountability function, preventing **freelancing,** and ensuring that provisions are made for responders' rehabilitation.

ACCOUNTABILITY

Personnel accountability is a critical element of the IMS and of the safety program. Accountability occurs by using several layers of supervision in various geographic areas or functional areas. Each supervisor is responsible for his or her operating crews. An accountability officer should be assigned within the IMS system, or, on smaller incidents, can be handled by the incident commander or safety officer.

As with the IMS, there are several variations in accountability systems in use. Some use the passport system where the crew's names are placed on small Velcro tags and placed on a small card that is used at the incident scene as a passport (see Figure 5-11). Others use two-dimensional bar codes and computers. Yet another system uses a card and clip that are attached to a large ring carried in the apparatus. Systems can be purchased commercially or designed locally, but the accountability system must meet the following objectives:

- Account for the exact location of all individuals at an emergency scene at any given moment in time
- Provide for expanding to meet the needs of the incident
- Be adaptable to the IMS in use
- Ensure that all individuals are checked into the system at the onset of the incident

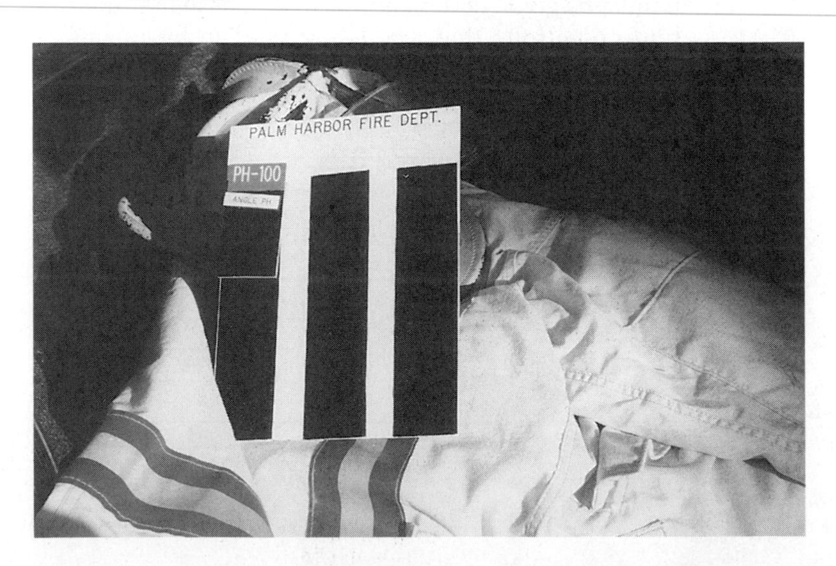

Figure 5-11 *The typical passport and accountability board.*

- Provide for visual recognition of participation
- Provide for points of entry into the hazard zone
- May also include medical data and the individual's training data

■ Note
Locally adopted accountability procedures are a strict requirement and oblige participation by everyone at all levels and at all incidents.

Locally adopted accountability procedures are a strict requirement and oblige participation by everyone at all levels and at all incidents. For an accountability system to be effective, it must have support by senior management of the organization. Each level has certain responsibilities to the system. Individual responders are responsible to ensure that their name tags are in the proper location on the apparatus and at the scene they are responsible to stay in direct contact with their assigned crew. *No freelancing!* Officers are responsible for knowing the location of each person assigned to them and to stay within their assigned work area. Sector or group officers are responsible for the accountability within the work area. Accountability officers are responsible for the accountability of any area of the incident from a specific point of entry. Incident command is responsible for putting the accountability function into the IMS and to provide a means for accounting for every individual at the scene. An accountability division/sector/group at a large incident may determine the points of entry, ensure communication with accountability officers, and provide the overall coordination of accountability at a specific incident.

personnel accountability reports (PAR)
verbal or visual reports to the incident commander or to the accountability officer regarding the status of operating crews; should occur at specific time intervals or after certain tasks have been completed

The accountability system requires **personnel accountability reports** (PARs). These PARs require the incident commander or accountability officer to check the status of crews and ensure that all personnel are accounted for during certain regular intervals or benchmarks during the incident. These benchmarks should be defined and may be related to time of activity. Examples of benchmarks include:

- A fixed time, generally every 30 minutes
- After primary search
- After fire is under control
- After a switch in strategy (offensive to defensive)
- A significant event such as a collapse, flashover, back draft
- After any report of a missing firefighter

RAPID INTERVENTION COMPANIES

rapid intervention companies (RIC)
assignment of a group of rescuers with the sole purpose of rapid deployment to reports of operating personnel in trouble or missing

Regardless of the effectiveness of the IMS in place, or the greatest of accountability systems, nothing is gained if personnel are not available to respond to assist when an unexpected event occurs at the emergency scene. The IMS and accountability system can provide the incident commander with the information that a crew is missing, but to intervene, the incident commander must have the resources close by to handle the situation. This is the concept of **rapid intervention companies (RIC)** or teams. Although called by many other names, the RIC is a group of firefighters, be it two, three, four, or more, that are fully equipped and have complete PPE on who stand by near the emergency scene and wait for something to happen. When something does happen, this crew is ready to respond and assist, whether it is a lost or trapped firefighter, or injured firefighters requiring assistance. The RIC would meet the requirements of the two in–two out rule introduced in Chapter 2.

Rapid intervention crews can be implemented in a number of ways. Some departments dispatch an extra unit to working fires to function as this team; others dispatch a company on the first alarm, for example, the fourth due engine might function as the RIC. Some use rescue or advanced life support units, others have designated teams such as special operations that respond to all structure fires.

The number of RIC necessary at an emergency incident can vary with the complexity of the incident. A room and contents house fire would probably only require a crew of two or three, while a large warehouse complex may require a RIC on each side of the building simply because of distance.

■ Note

Once assigned to rapid intervention, the companies should not be assigned other tasks and should have direct communication to the incident commander.

Once assigned to rapid intervention, the companies should not be assigned other tasks and should have direct communication to the incident commander. On arrival, the team should size up the building and the operation and try to anticipate what could happen. It should determine possible entry and exit points. For example, in a two-story garden apartment building, the RIC may want to have a 24-foot ladder with them for rapid access to the second floor.

This RIC concept has widespread applicability, not only to fire emergencies but other incidents as well. The incident commander must provide for a RIC early in this incident and throughout the duration.

REHABILITATION

The physical and mental demands placed on responders coupled with the environmental dangers of extreme heat and cold will have an adverse effect on the responder from a safety perspective. Figures 5-12 and 5-13 show the heat stress index and the wind chill index, respectively. Crews that have not been provided with adequate rest and **rehabilitation** during a fire or other emergency operation are at increased risk for illness and injury.

rehabilitation
group of activities that ensures responders' health and safety at an incident scene; may include rest, medical surveillance, hydration, and nourishment

Rehabilitation on the emergency scene is an essential element of the IMS (see Figure 5-14). This need for rehabilitation is also cited in a number of the national standards relating to safety and fire scene operations.

The development of an incident rehabilitation program has minimal impact on an organization and therefore should be achievable. While systems can be very different, the basic plan should provide for the following:

■ Note

Rehabilitation on the emergency scene is an essential element of the IMS.

- Establishment of a rehabilitation sector/group within the IMS
- Hydration
- Nourishment
- Rest, recovery
- Medical evaluation

TEMPERATURE °F	RELATIVE HUMIDITY 10%	20%	30%	40%	50%	60%	70%	80%	90%
104	98	104	110	120	132				
102	97	101	108	117	125				
100	95	99	105	110	120	132			
98	93	97	101	106	110	125			
96	91	95	98	104	108	120	128		
94	89	93	95	100	105	111	122		
92	87	90	92	96	100	106	115	122	
90	85	88	90	92	96	100	106	114	122
88	82	86	87	89	93	95	100	106	115
86	80	84	85	87	90	92	96	100	109
84	78	81	83	85	86	89	91	95	99
82	77	79	80	81	84	86	89	91	95
80	75	77	78	79	81	83	85	86	89
78	72	75	77	78	79	80	81	83	85
76	70	72	75	76	77	77	77	78	79
74	68	70	73	74	75	75	75	76	77

NOTE: Add 10°F when protective clothing is worn and add 10°F when in direct sunlight.

HUMITURE °F	DANGER CATEGORY	INJURY THREAT
BELOW 60°	NONE	LITTLE OR NO DANGER UNDER NORMAL CIRCUMSTANCES
80° - 90°	CAUTION	FATIGUE POSSIBLE IF EXPOSURE IS PROLONGED AND THERE IS PHYSICAL ACTIVITY
90° - 105°	EXTREME CAUTION	HEAT CRAMPS AND HEAT EXHAUSTION POSSIBLE IF EXPOSURE IS PROLONGED AND THERE IS PHYSICAL ACTIVITY
105° - 130°	DANGER	HEAT CRAMPS OR EXHAUSTION LIKELY, HEAT STROKE POSSIBLE IF EXPOSURE IS PROLONGED AND THERE IS PHYSICAL ACTIVITY
ABOVE 130°	EXTREME DANGER	HEAT STROKE IMMINENT!

Figure 5-12 *Heat stress index. Courtesy United States Fire Administration.*

- Accountability while in the rehabilitation sector
- Supplies, shelter, and number of people needed to operate the rehabilitation area

Procedures should define the responsibilities within the IMS for the incident commander, supervisors, and personnel. Further, the procedures should provide guidelines for location, site characteristics, site designations, and resources

WIND SPEED (MPH)	TEMPERATURE °F 45	40	35	30	25	20	15	10	5	0	-5	-10	-15
5	43	37	32	27	22	16	11	6	0	-5	-10	-15	-21
10	34	28	22	16	10	3	-3	-9	-15	-22	-27	-34	-40
15	29	23	16	9	2	-5	-11	-18	-25	-31	-38	-45	-51
20	26	19	12	4	-3	-10	-17	-24	-31	-39	-46	-53	-60
25	23	16	8	1	-7	-15	-22	-29	-36	-44	-51	-59	-66
30	21	13	6	-2	-10	-18	-25	-33	-41	-49	-56	-64	-71
35	20	12	4	-4	-12	-20	-27	-35	-43	-52	-58	-67	-75
40	19	11	3	-5	-13	-21	-29	-37	-45	-53	-60	-69	-76
45	18	10	2	-6	-14	-22	-30	-38	-46	-54	-62	-70	-78

A B C

	WIND CHILL TEMPERATURE °F	DANGER
A	ABOVE 25°F	LITTLE DANGER FOR PROPERLY CLOTHED PERSON
B	25°F / 75°F	INCREASING DANGER, FLESH MAY FREEZE
C	BELOW 75°F	GREAT DANGER, FLESH MAY FREEZE IN 30 SECONDS

Figure 5-13 *Wind chill index.*

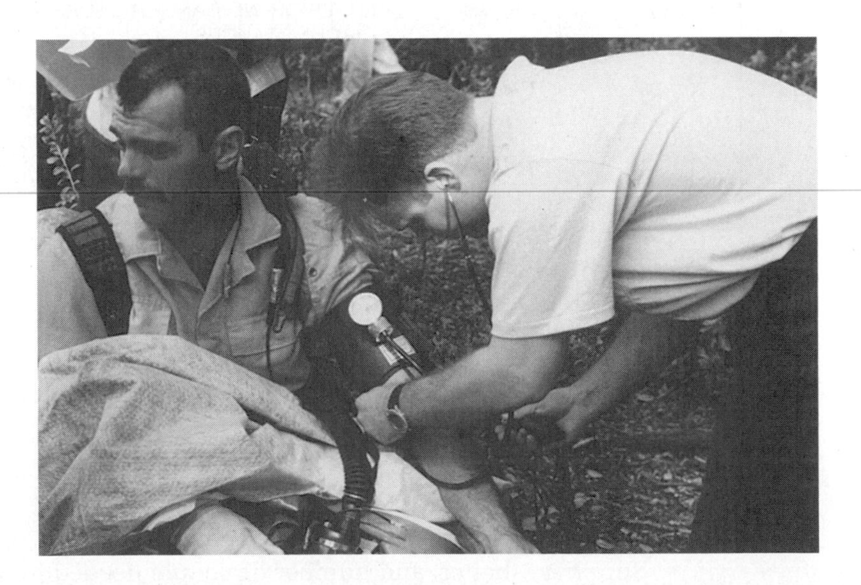

Figure 5-14 *A firefighter beginning the rehab process.*

■ Note

Remember that the medical information from a rehabilitation sector or group may be deemed confidential medical records. Check with your local legal advisor.

required. Records should be kept on all individuals entering the rehabilitation area and medical evaluation documented. Remember that the medical information from a rehabilitation sector/group may be deemed confidential medical records. Check with your local legal advisor.

Summary

Fireground activities account for the highest percentages of firefighter deaths and injuries of any other specific type of duty. It is essential that the safety and health program address safety issues relating to these types of incidents. This requires an understanding on behalf of the safety program manager regarding operational priorities and their application to personnel safety. The operational priorities are always life safety (including the responders), incident stabilization, and property conservation. Understanding the priorities helps to perform risk analysis, which should be based on the premise, we will risk much to save much, we will risk little to save little, and we will risk nothing to save what is already lost.

Hazards faced by responders at fire incidents vary by the type and complexity of the incident. To minimize exposure to the hazards, the safety program must include components relating to personal protective equipment, require the use of an incident management system, provide for an accountability system, ensure rapid intervention companies are deployed, and have a rehabilitation process. Personal protective equipment concerns are based on design and purchase, use, and care, maintenance, and inspections.

The type of incident management system may vary provided that it meets certain criteria. Within the incident management system, there must be a safety officer with the proper authority and responsibility for incident scene safety. Accountability systems provide for a strict accounting of all personnel throughout the incident. When an unexpected situation occurs, the incident commander must have a team available to respond and assist. This role belongs to the rapid intervention company. Prolonged physical and mental stress can adversely affect emergency responders at the incident scene. Proper procedures for rehabilitation must be part of the incident management plan.

Concluding Thought: Having a practical system in place to handle fire-based incidents, including consideration for personnel and equipment, and having the processes in place for safety mitigation will result in safer fireground operations.

Review Questions

1. Property conservation is the first incident priority.
   A. True
   B. False
2. Explain how the first incident priority can have a positive impact on firefighter safety.
3. List three hazards commonly found at building fires.

4. Which of the following is not an incident priority?
   A. Preplanning
   B. Incident stabilization
   C. Property conservation
   D. Life safety
5. List the ten firefighting orders as related to wildland fires.
6. List the components of an effective PPE ensemble for firefighting.
7. List five characteristics of an effective accountability system.
8. List the positions in the command staff and general staff of an IMS.
9. What three authorities must an incident safety officer be given to be effective on the fireground?
10. How often should firefighting personnel inspect their PPE?
    A. Annually
    B. Weekly
    C. Monthly
    D. Daily and after each use

Activities

1. Review your department's PPE inspection program. What changes do you feel are necessary to make it more effective?
2. Review your department's procedures for the assignment of rapid intervention companies. Do you use them? Are they effective? What improvements can be made?
3. After performing the risk identification process described in Chapter 3, make a matrix for your jurisdiction relating to risks found at building fires.
4. Review your department's rehabilitation SOP. Does it meet the requirements discussed in the chapter?

Safety at the Medical or Rescue Emergency

Learning Objectives

Upon completion of this chapter, you should be able to:

- Discuss the hazards faced by responders at emergency medical or rescue incidents.
- Explain the methods of minimizing and preventing injuries associated with hazards found at emergency medical incidents.
- List various requirements and uses of commonly used personal protective equipment.
- Explain the requirements for infection control.
- Discuss procedures that can be used to meet the requirements of infection control.
- Discuss the systems used for scene accountability and incident management.

CASE REVIEW

It was a routine maternity call. The medic unit responded with a crew of two to the apartment and was met by relatives anxious that the day for the newest member of the family to arrive had come. Paramedics went in and found the patient in a back bedroom. The patient was in her early twenties and in her ninth month of a first pregnancy. It seemed routine enough, get vitals, time contractions, start an intravenous (IV) line, and transport. The crew began their work.

As the treatment of the patient began, the crew heard some commotion in the front room. Knowing the large number of family members and their excitement level, they did not think anything of the commotion. Soon a man appeared in the doorway to the bedroom shouting to leave his wife alone. He lurched forward with an 8-inch butcher knife and stabbed one of the crew members in the arm. Police units were summoned and arrived to arrest the man. A second ambulance took the mother-to-be to the hospital. The paramedic who was stabbed was transported by his own medic unit. The injury was minor and, although the medic lost some work, he did recover. It could have been a lot worse.

Emergency medical services (EMS) alarms comprise the majority of alarms that emergency responders face. Although the calls become routine, the potential of risk to responders is high.

INTRODUCTION

Whereas Chapter 5 focused on fire scene and incident safety, this chapter focuses on the type of incidents that most responders face most of the time. It is not uncommon to find rural, suburban, and urban departments involved in EMS, responding to EMS incidents 80% or more of the time. Therefore EMS safety is a critical and integral part of any department's safety and health program.

HAZARDS FACED BY RESPONDERS

What makes the emergency medical incident scene so dangerous? There are many factors to consider when answering this question. One such answer might be that, like fire officers, EMS crews must make critical, time-sensitive decisions while on EMS alarms: the safest and quickest route to the scene, drug dosages, treatment regimes, the most appropriate form of transportation and the receiving facility, and choosing the best interpersonal approach to the patient. Obviously, the decisions occupy a great deal of the responders' thought processes, however the responder must factor in safety in all of these decisions. A good example might be the response to a multiple car accident on a divided highway that runs from north to south. The accident may be in the northbound lanes, but the ambulance is approaching from the south. The quickest means of reaching the patient would be to drop off a responder and have him or her cross the traffic to reach the victim while the driver went to the next exit and came around. A safer approach, considering, of course, the distance to the exit, would be for both responders to continue to the exit and turn around. This might add an additional few minutes to arrival on the scene, which must be weighed against the threat to the responder in crossing the road. Which approach would you take?

Another question is the validity or accuracy of information from dispatch. Have you ever heard of a response to a man down, which turned out to be a shooting? Not usually the dispatcher's fault, this breakdown of information commonly occurs at the calling party's end. With the increase in cellular phone use, alarms are received sooner, but often the caller is a passerby and does not know a lot about the emergency.

■ Note
With the increase in cellular phone use, alarms are received sooner, but often the caller is a passerby and does not know a lot about the emergency.

The responders also face safety issues when dealing with the patients, family, and bystanders. Remember EMS events, as with all emergency incidents, are emotional events; people at the scene may not be thinking rationally and they expect immediate action from the responders. Try to tell the frantic father of child who has been electrocuted about the safety issues and the need to cut the power. Or try to tell the outraged family member of the shooting victim that you must wait for the police until you can enter the house.

Like some fire-related incidents, responders to routine emergency medical calls often develop a complacence toward the type of calls (see Figure 6-1). How many chest pains calls have you responded to? Do you approach the patient with less caution now than you did your first days on the job? How many helicopter landings? Do you take the same precautions now as you did for your first landing zone setup? The answer very well maybe yes, and that is the correct approach. However, the reality is that because you have attended numerous similar incidents and everything has gone well, you may let your guard down. This chapter reminds, reinforces, and, in some cases, gives a new look at EMS incidents and the threats to responders.

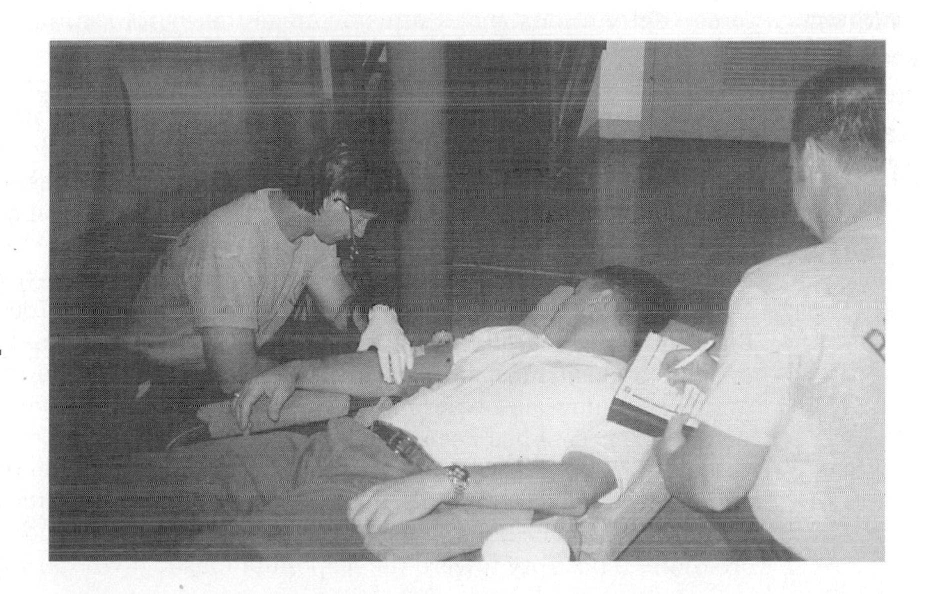

Figure 6-1 *EMS incidents are a safety concern for responders regardless of how routine the incident may seem.*

General Hazards for All EMS Incidents

Once the responder has arrived on the scene, the incident can be viewed in four phases:

- Gaining access
- Approaching and first contact with the patient
- Providing care
- Packaging and placing the patient in the transport unit

There are obviously additional issues associated with a medical incident, such as response to the scene, but they are addressed elsewhere in this text.

The first phase of the incident is to gain access to the location, whether it be a home, apartment, hotel, restaurant, or vehicle. Upon arriving at the incident scene, park and secure your vehicle in a safe, but accessible, location.

One consideration on gaining access are those situations in which forced entry is required. There are a number of concerns, the least of which is security after the patient is removed. Forcible entry is necessary when a patient is unable to come to the door. Law enforcement should be requested to respond to all forcible entry situations and where possible the forcible entry should wait until their arrival. Forcible entry is in itself a safety hazard. By definition, forcible entry requires that something will be broken, such as glass or door frames. Responders should have on hand the necessary personal protective equipment and the proper tools when attempting forcible entry. If the equipment and tools are not available, then additional resources should be called. Before you force entry, consider that another opening may be unlocked. Many times a second-floor window or balcony door may be left open or unlocked, requiring only the use of a ladder. Before any forced entry occurs, make sure you are at the right location.

Note

Before any forced entry occurs, make sure you are at the right location.

In most cases forced entry is not required and the responder can enter the location as any other person would. But there are still considerations about how you enter and what you see when you do. The responder should always carry a flashlight at night and consider it during the day at some locations. In the event the patient is in a basement or other dimly lit area, the flashlight will be necessary so it is best to carry it always. When looking around on entry, the same considerations should be noted as on arrival outside. Look for access to the patient and consider transport options, poorly repaired stairs, rugs, tripping hazards, and animals. Animals present in the house should prompt the responder to ask that the animal be secured in another room. Although the animal may seem friendly or small enough, most animals are very protective of their owners and an attempt to touch or move an owner may induce an unwanted response from the animal.

Once access has been made and the patient located, phase two, approach and first contact, begins (see Figure 6-2). There are a number of things to consider when first contact is made, most of which deal with interpersonal skills. An approach to a conscious patient should start with an introduction and why you are there. If hostility toward the responder exists, it will be apparent at this time.

Figure 6-2 *EMS responders should be alert for safety issues when first contact is made with patient.*

Be aware of threats or statements that indicate that the patient may not want your help. Again this may be an indication that police assistance is needed and a retreat necessary. Note the patient's hands and surroundings and make sure nothing is present that might be used to hurt you or cause injury as you work around the patient. Glass on the floor is a good example, considering you will probably kneel down to take vital signs. Always approach cautiously and move items that may pose problems in the next phases of the incident out of the way.

Unconscious patients should be approached with caution as well. Is it possible that you might be waking up a patient who just paid $100 for a high and passed out? Or is it possible that the patient is just sleeping and could react violently when awakened? Again, a cautious approach and being prepared to back off is required.

■ Note
During first contact, the responder must use a variety of interpersonal skills.

During first contact, the responder must use a variety of interpersonal skills. Tone of voice is important: Commanding respect would be necessary in a nightclub, while soft and gentle might be the order for a small child at home. An understanding of personal space is important, remembering that in the space in which the responder works, taking vital signs, holding spinal support, and the like involve touching the patient. It is generally accepted in the United States that intimate space is 18 inches around a person. Therefore, any unfamiliar responder immediately invades this space and the patient may experience a feeling of uneasiness, leading to an unwanted reaction. It is, therefore, very important for the responder to develop trust early on.

After, and almost simultaneous with first contact is the providing of care phase of the response. In this phase the responder is again faced with several safety challenges, one of which is exposure to communicable diseases, which is discussed under Infection Control. Others include injury during moving the patient, needle sticks, and contact with objects around the patient. Usually if the necessary precautions have been taken in the first two phases and the situation does not involve communicable diseases or hostility, this phase can proceed relatively safely.

The final phase of the EMS incident to be discussed in this chapter is packaging and placing the patient in the transport unit. Now is the time when many of the mental notes that were made on arrival become important. Are the stairs safe to carry the patient? If not, what other means are available? Consider that stairs not only have to support the weight of the responders, but also the patient being carried. Perhaps a fire department ladder truck with a bucket might be a safer alternative. Were there any obstructions or tripping hazards, either inside or outside, on the approach? Are rugs and carpets secured down? If not, roll them up to create a clear path. Finally, are enough people present to make the move safely, considering outside weather conditions, the patient's weight and location, and available equipment.

Hazards Associated with Vehicle Accidents and Incidents on Roadways

Operations associated with incidents on roadways pose distinct safety hazards, as compared to incidents inside a structure. Responders to incidents on roadways have much less control over their surroundings (see Figure 6-3). Traffic, weather, and crowd control are also concerns.

Upon arrival at this type of scene, the first order of tactical concern is where to safely locate the response vehicle. Consideration should be given to parking the unit in a manner that will provide the best level of protection for the responders.

Figure 6-3 *Motor vehicle accidents put the responder at risk from traffic.*

Other considerations are which location will best facilitate patient care and transport, which location will provide the best departure, and whether other emergency vehicles will be arriving. Of course, safety is the primary concern when considering vehicle placement.

There are differing thoughts on vehicle placement from the safety standpoint. One thought is to place the emergency vehicle at the end of the accident scene closest to oncoming traffic. The justification here is that the emergency warning lights give oncoming drivers warning, the scene is illuminated by the vehicle headlights, and there is a physical barrier between oncoming traffic and responders. However, once the patient and responders move into the unit, they are endangered by being closest to oncoming traffic. A better situation is one in which multiple units respond, for example an ambulance and a rescue squad or engine. In this case, the transport unit can be parked on the other side of the incident scene and the large vehicle on the end nearest oncoming traffic (see Figure 6-4). Locating the unit also depends on other hazards that could be present at the scene, such as fuel spillage, downed power lines, or hazardous cargo. Always try to avoid parking the vehicle in a manner that requires emergency responders to cross traffic to get to and from patients.

Traffic always presents a hazard to responders to roadway incidents. Several safety principles should be applied. Create a buffer zone by using an emergency vehicle as described in the previous paragraph. Creating a nontraffic lane next to the incident site may be another option. Do not be afraid to close the road

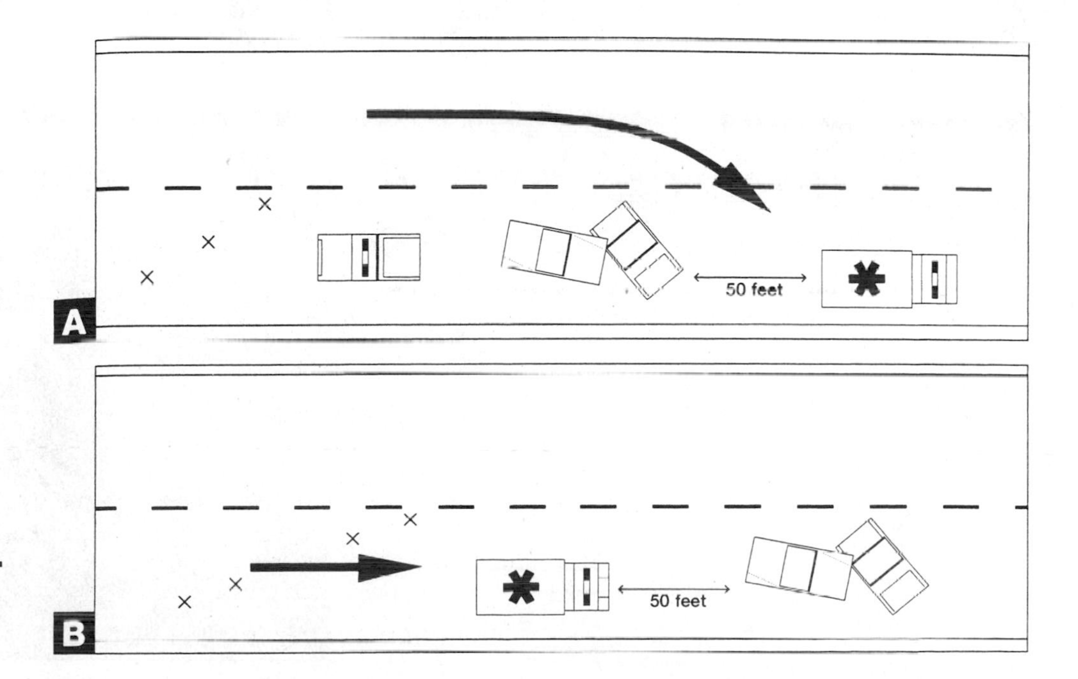

Figure 6-4 *Two examples of traffic control options.*

to protect the responders. This can be accomplished through the use of orange traffic cones and law enforcement personnel. Finally, wear protective clothing or a vest with reflective trim regardless of whether the incident is in the daytime or after dark.

Other hazards related to roadway incidents, specifically vehicle accidents, are those hazards associated with vehicle rescue or extrication. Vehicle rescue is comprised of action to gain access, disentangle, and remove patients trapped in vehicles. Most often this is accomplished using heavy hydraulic equipment for cutting and spreading metal and car components (see Figure 6-5). Aside from the hazards of the accident itself, using this equipment produces additional hazards, such as flying glass and metal, potential catch of rescuers' fingers and hands between the vehicle and the cutting or spreading tool, and fire hazards as a result of using powered tools. Any time vehicle rescue operations are being performed, an incident management system (IMS) should be established, a safety officer assigned, full protective clothing designed for thermal and abrasion resistance should be worn, and fire suppression capabilities should be present (see Figure 6-6).

Figure 6-5 *Examples of heavy extrication equipment.*

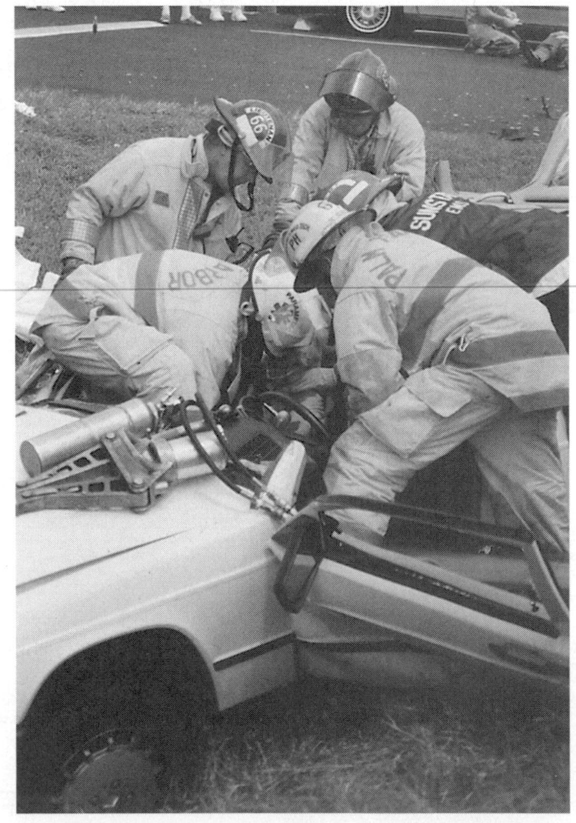

Figure 6-6 *A vehicle extrication in progress.*

In addition to vehicle accidents, other EMS-needed incidents occur on roadways. Injuries from violence including shooting, stabbing, and assault, are too prevalent in the country today. Responders to these incidents should be aware of these situations and rely on law enforcement support and scene control. The hazards associated with these incidents are obvious. At a drive-by shooting, for example, who is to say the shooters will not drive by again after responders have arrived.

Environmental hazards are also a concern in roadway situations. Extreme cold or hot, rain, snow, sleet, and the duration of the operation can impact the responder's safety. Rehabilitation considerations similar to those presented in Chapter 5 should be a part of the response plan.

Chapter 4 discussed the importance of interagency preplanning and interaction. This is very important for roadway incidents. Law enforcement is needed for crowd and traffic control and to ensure the securing of the scene following incidents involving violence. Preplanning the necessary equipment, resources, and expectations of both agencies is essential to maximize safety on the scene.

INFECTION CONTROL

Infection control is a requirement for emergency responders to EMS incidents. Chapter 2 presented several federal regulations that govern these programs. Infection control is essential, not only for the responders' safety, but also to maximize the safety of the patients they treat. The safety program manager may or may not be the department's infection control officer. Regardless of who it is, the law requires the organization to assign an infection control officer to handle exposures, follow-up care, and record keeping. The organization must also adopt an infection control plan to prevent the transmission of diseases.

The infection control plan is based on engineering controls, personal protective equipment, and education. Engineering controls can be based on risk management principles. For example, the risk of needle sticks can be reduced or eliminated by changing the practice of resheathing needles and just placing them in a sharps container. Personal protective clothing is discussed in the next section. Education requires training employees in the types of infectious diseases, routes of transmission, and prevention strategies.

■ Note

Responders are exposed to a number of diseases, some minor, some serious, some even fatal.

Responders are exposed to a number of diseases, some minor, some serious, some even fatal. An exposure is not the single mechanism needed to acquire the illness; several factors must interplay for an infection to occur. As with injury prevention, the infection can be prevented by interrupting the disease process at any of these points:

Dose: the number of live organisms present

Virulence: the strength of the infecting organism

Host Resistance: the ability of the host to resist the effects of the infectious organisms

Route of Exposure: airborne, bloodborne, or foodborne

Means of Transmission: a way for the organism to gain entry into the host—injection, inhalation, absorption, or ingestion

An infection control manual can provide information regarding the listed factors for specific diseases. One of the best ways to intervene or interrupt the disease process is to prevent the means of transmission. Diseases are transmitted via direct or indirect contact, an intermediate host, or other vehicles.

Direct contact transmission occurs when physical contact occurs between the responders and the infected person. Indirect contact occurs when an infected person contacts an object and then the responder contacts the same object. Of course, in the case of bloodborne disease, there would have to be blood on the object. Another means of indirect contact is droplet contact, which occurs when an infected person coughs or sneezes and airborne droplets are produced. These microscopic droplets can travel a great distance.

Examples of transmission via an intermediate host would be diseases spread by ticks, mosquitoes, flies, or fleas. Disease spread through water and food are examples of transmission via other vehicles.

■ Note
Responders should take advantage of vaccinations and boosters for diseases when available.

As described in Chapter 4 in the section on employee wellness and fitness, responders should take advantage of and be provided with boosters and vaccinations for diseases when available. Responders should also be trained in personal hygiene and equipment cleanup procedures. The response station must have clearly marked areas for cleaning contaminated equipment. In terms of personal hygiene, hand washing is particularly important. Since hand washing facilities are not always available, waterless gels that are disinfectants should be carried on the emergency vehicle.

PERSONAL PROTECTIVE EQUIPMENT

Personal protective equipment (PPE) for responders to EMS incidents is specialized and differs from that of firefighting PPE. EMS responders who are cross trained as firefighters are issued firefighting PPE that is appropriate for most EMS incidents, such as vehicle rescue, because firefighting PPE provides thermal and abrasion resistance. Because firefighting PPE was described in Chapter 5, this section focuses on PPE particular to EMS emergencies that are unrelated to entrapment or rescue.

The requirements for PPE for body substance isolation (BSI) are defined in the respective regulations. The EMS providers often must make a decision as to what PPE is necessary at particular incidents. Procedures should be developed to give guidance in PPE selection for tasks at the emergency scene. Table 6-1 is an example of a matrix that should be included as part of an infection control SOP. It is *always* the rescuer's option to do more than the minimum requires based on the circumstance. From head to toe, the EMS responder should have PPE ade-

Table 6-1 *PPE matrix.*

Task or Activity	Disposable Gloves	Gown	Mask	Protective Eyewear
Bleeding control with spurting blood	X	X	X	X
Bleeding control with minimal bleeding	X			
Emergency childbirth	X	X	X	X
Blood drawing	X			X
Starting an intravenous (IV) line	X			X
Endotracheal intubation, esophageal obturator use	X		X	X
Oral/Nasal suctioning, manually cleaning airway	X		X	X
Handling and cleaning instruments with possible microbial contamination	X	X	X	X
Measuring blood pressure	X			
Giving an injection	X			
Measuring temperature				
Cleaning back of ambulance after a medical alarm	X	X	X	X

quate for the hazards to be faced. Table 6-2 charts PPE for various body areas. These are recommendations and must be geared to local conditions.

EMS PPE should be cared for, disposed of, and inspected according to the manufacturer's recommendations. The inspection program should be similar to that of fire PPE, however, much of the EMS PPE is disposable. The same record keeping should be required for EMS PPE as for firefighting PPE.

INCIDENT MANAGEMENT SYSTEMS

Like fire incidents, emergency medical/rescue incidents must be managed with an IMS. This is not to say that responders must break out the command board for every heart attack call, but in reality IMS does occur on each incident. It is just

Table 6-2 *Personal protective equipment for body areas*

Body Area	PPE Item	Remarks
Head	Helmet	If not provided as firefighting PPE
Head	Eye protection	To protect from objects and body fluid intrusion
Head	Hearing protection	For prolonged rescue in noisy environment
Head	Mask	To prevent airborne transmission or fluid from entering the mouth
Torso	Abrasion-resistant jacket	If not provided as firefighting PPE
Torso	Gown	To prevent fluid transmission
Arms	Protective sleeves	To prevent fluid transmission
Hands	Abrasion-resistant gloves	If not provided as firefighting PPE
Hands	Fluid-resistant gloves	
Legs	Abrasion-resistant pants	If not provided as firefighting PPE
Legs	Gown	To prevent fluid transmission
Feet	Shoes hard sole, steel toe	
Feet	Shoe covering	To resist fluid penetration
Body	Body armor	Dependent on local conditions and risks

that the senior EMS responder assumes many of the roles in the command staff. On multiple unit or multiple jurisdictional response, IMS is as much required on the EMS incident as it is on the fire or hazardous materials scene.

For an EMS incident, the IMS can be adapted. The three command staff and four general staff functions remain the same. Responsibilities within the command structure only differ depending on the type of incident. At a multicasualty bus accident without entrapment, there may be a group/sector for triage, treatment, and transport. At a single vehicle in a pole with entrapment, there may be groups or sectors for hazard control, extrication, and treatment. The system is adaptable to any emergency. One common factor is the incident safety officer. Although very dependent on the situation, the safety officer should be assigned any time responders are performing hazardous operations such as extrication. On a routine medical incident the senior person or officer must assume all of the command functions including safety. In this role, the safety officer can ensure and require that proper PPE is being worn for the particular type of incident. Figure 6-7 details an expanded IMS at a medical/rescue incident.

■ Note

On a routine medical incident, the senior person or officer must assume all of the command functions including safety.

An accountability system for EMS responders must be adhered to as with fire incidents. In some jurisdictions, some parts of EMS response are handled by private or third service public agencies. When these agencies respond and are required to work at the incident they must be part of and work within the IMS structure and the accountability system. The safety program manager for the EMS provider should ensure that all EMS personnel have the proper training and equipment to work within these two systems. This also can be part of interagency coordination.

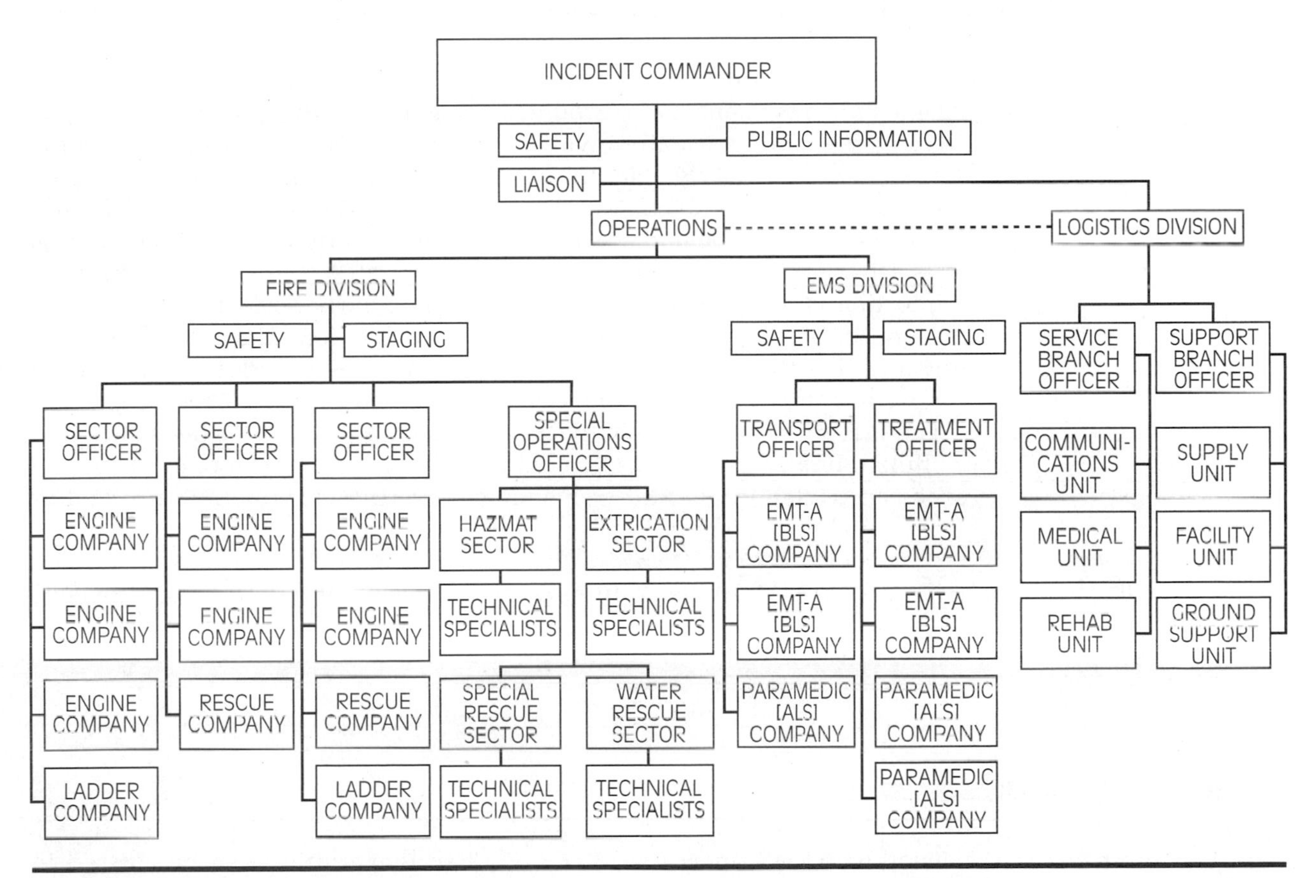

Figure 6-7 *An expanded IMS including EMS functions. (Courtesy United States Fire Administration, Emmitsburg, MD.)*

Summary

For most jurisdictions EMS emergencies comprise the highest majority of incidents. The hazards faced by responders are increased by uncertainties and the need to interact closely with the patients. Medical or rescue incidents can be generally divided into those that occur inside and those that occur outside. Inside structure incidents, often viewed as routine can cause injuries. A step-by-step approach and an awareness of surroundings can reduce this potential.

Outside incidents usually involve vehicle accidents. Hazards such as the vehicle, the fuel for the vehicle, traffic, and the environment cause additional problems for the responders. Anticipating these hazards through training on proper vehicle placement and control of the scene can reduce exposure to these hazards. Utilizing law enforcement for crowd and traffic control is also required.

Infection control is a relatively new addition to the hazards involved in medical response. Through the use of engineering, education, and PPE, the risk of acquiring an infectious disease can be reduced. The use of PPE is both required and an excellent means to interrupt the disease transmission process.

The IMS must be adaptable to EMS events and provide for the same command functions and structure as that of fires. The system must include a method for scene accountability of EMS responders, particularly if the EMS responders are from a different agency.

Concluding Thought: EMS and fire events have many of the same requirements for success. Engineering, education, and PPE are three requirements essential to the safety program.

Review Questions

1. List four hazards associated with EMS emergencies that occur indoors.
2. List three hazards that would be present outside on a highway at a vehicle accident.
3. Which of the following practices should a responder to a vehicle accident NOT take?
   A. Create a buffer zone
   B. Create a nontraffic lane
   C. Park so that traffic must be crossed to access patients
   D. Wear reflective vests
4. What three functions might be in the IMS for a multicasualty incident?
5. List the five factors that must interplay for an infection to occur.
6. Which of the factors in Question 5 is easiest to control?

7. In terms of the spread of infection, compare the terms *direct* and *indirect contact*.
8. What are the three components involved in vehicle rescue?
9. What PPE components might also be part of firefighting PPE?
10. List five tasks or activities that should be included in a required PPE matrix.

Activities

1. Review your department's procedures and practices for responding to vehicle accidents. Do these apply the safety principles as discussed in this chapter?
2. Review your department's infection control plan and compare it to the federal requirements. Do you have an infection control officer, and does this officer regularly respond to incidents? Does the plan provide for engineering controls, education, and appropriate levels of PPE?

Safety at Specialized Incidents

Learning Objectives

Upon completion of this chapter, you should be able to:

- Describe the safety issues related to hazardous materials incident response.
- Describe the safety issues relating to technical rescue operations.
- Explain safe procedures to be used during helicopter landing zone operations.
- List specific safety issues relating to operations at civil disturbances.
- List specific safety concerns when responding to terrorism events.

CASE REVIEW

One relatively quiet Sunday afternoon, the crew of a medium-sized Florida fire department received a telephone call from a concerned emergency room nurse from a nearby hospital. She reported that a patient had been brought into the emergency room via a private car. The patient complained of shortness of breath, but the cause was unknown. As the patient was elderly, normal care for shortness of breath was established. During the routine physical exam, burns that appeared to be chemical burns were noted in the patient's mouth. With the patient's level of consciousness in question, the nurse called the fire department to ask about possible hazardous materials incidents that day. The fire department had not had any hazmat-related calls that day, but took down the patient's address and assured the nurse that they would look into it.

The address was in a local mobile home park. A single engine and battalion chief responded to check the address. Prior to arrival, the crew discussed an approach strategy. Because the potential existed for some type of chemical at the location, a very cautious approach was undertaken. Upon arrival, units staged upward, donned protective clothing, and, with monitor in hand, approached the address. At the front door the vapor monitor went into alarm mode and the crews backed off, setting up a perimeter, and evacuating several mobile homes nearby.

The fire department discovered a lethal amount of ammonia in the mobile home. The incident lasted about 16 hours until the ammonia leak could be bled off. The outcome was good; the patient did well, and the responders operated in a safe fashion.

But, what if the patient had remained at home and called 911 and reported himself to be short of breath. How would the crew have responded? Probably the same way they had responded to hundreds of other shortness of breath incidents. However, in this case they may have entered a mobile home and been exposed to high amounts of a hazardous substance. Responders must always be aware of potential hazards, even when responding to seemingly routine incidents.

INTRODUCTION

The discussion of safety issues at emergency scenes would not be complete without a look at safety and special operations. These particular types of incidents include hazardous materials response, confined space rescue, high-angle rescue, helicopter landing zone operations, water rescue operations, and operations at civil disturbances.

These specialized incidents are addressed in a separate chapter because some departments may not be involved in all of these operations. Some departments may not be involved in any of them. The local conditions and the services that a particular department provides dictate what safety measures and procedures must be in place. Each type of incident is discussed and basic safety concerns presented.

HAZARDOUS MATERIALS INCIDENTS

Hazardous materials response is routine to many departments. What differs significantly is the level of response. Some departments may have full hazardous

■ Note
As part of the overall safety and health program, the safety program manager must determine what level of response the department provides and compare that level to the training level of the responders.

materials teams and provide complete response services including offensive action to mitigate the incident. Other departments may respond at the first response level and provide defensive operations only, relying on a regional hazardous materials team for offensive action and mitigation. The response to hazardous materials incidents is governed by Occupational Safety and Health Administration (OSHA) and Environmental Protection Agency (EPA) regulations. These regulations require that certain safety measures be in effect before operations can begin. These requirements are described in Chapter 2 (see Figure 7-1).

As part of the overall safety and health program, the safety program manager must determine what level of response the department provides and compare that level to the training level of the responders. There are five levels of response training, as shown in Text Box 7-1.

Figure 7-1 *Hazardous materials response and operations produce unique safety concerns.*

TEXT BOX 7-1 FIVE LEVELS OF HAZMAT RESPONSE TRAINING.

First Responder Awareness: Responders at this level are likely to witness or discover a hazardous substance release and have been trained to initiate an emergency response sequence. At this level of training, these responders would be expected to take no further action. Typically, single-certified EMS responders and law enforcement officers fall into this category.

First Responder Operational: Responders trained to this level respond to hazardous materials incidents as part of the initial response for the purpose of protecting nearby persons, property, and the environment from the release. These responders are trained in defensive tactics without actually trying to stop the release. Their function is to contain the release from a safe distance, keep it from spreading, and protect exposures. First responder operational level requires at least 8 hours of training or demonstrated competency.

Hazardous Materials Technician: Responders trained to this level assume a more aggressive role than a first responder operational level. The hazardous materials technician approaches the point of release in order to plug, patch, or otherwise stop the release of a hazardous substance. Hazardous materials technicians must have received at least 24 hours of training equal to the first responder operational level and demonstrated competency in additional areas.

Hazardous Materials Specialist: Hazardous materials specialists are those individuals who respond to a hazardous materials scene to support technicians. Their duties parallel those of the hazardous material technicians, but they require a more direct knowledge of the various substances they need to contain. Hazardous materials specialists may also act as the liaison with federal, state, local, and other governmental authorities in regard to site activities. Hazardous materials specialists must have received at least 24 hours of training equal to the technician level, plus additional competencies as identified in the standard.

On-Scene Incident Commander: Incident commanders who will assume control of the incident beyond the first responder awareness level must receive at least 24 hours of training equal to the first responders operational level and in addition, have competency in command systems, response plans and options, and other hazards associated with hazardous materials.

Having identified the five levels of training, the program safety manager must review department procedures and practices to ensure that the responders are operating within the level that they have been trained and within the level of equipment available. A serious incongruence would occur if the department responded routinely to hazardous materials releases and plugged the leaks, but had only been trained to a first responder level.

There are numerous safety related issues regarding the hazardous materials incident. The risk factors increase with the level of service provided. Hazardous materials personal protective equipment (PPE) is a complete subject in itself.

Regulations require the use of incident command systems and backup teams with the same level of protection as the entry team. However, many of these issues relate only to departments that provide hazardous materials technicians and do offensive mitigation strategies. These safety measures should be addressed in the hazardous materials teams operating procedures.

One area of hazardous materials response that is applicable to all responders is the initial response and arrival. Procedures and training for this segment of the incident should focus on safety and the initial response. One method to use to maximize safety of the initial response is to remember the acronym RAID, whose letters stand for Recognize, Approach, Identify, and Decide. See Text Box 7-2.

TEXT BOX 7-2 THE RAID PROCESS.

Recognize. The recognition phase is when the responder recognizes that an incident might involve a hazardous substance. The recognition process begins with preplanning. Being aware through preplanning what hazards are present in a given response area helps to determine the precautions that will be taken at an incident at that specific location. Clues to the presence of hazardous materials also present themselves at the time of dispatch: The location, occupancy, and the name of the business can provide clues. For example, a report of person (s) down behind Jones Pool Chemicals should give the responders a pretty strong clue that hazardous materials may be involved.

Approach. During the approach phase, several things must be considered. If the responders suspect that a hazardous substance is involved, the approach should be cautious and from uphill and upwind if possible. During the approach, the responders should be looking for other clues, such as strange-colored smoke, vapor clouds, and employees running from the area. If necessary, park the response unit away from the scene until further analysis can be performed. In the Jones Pool Chemicals example, the crew should be alert for vapor clouds or smells such as chlorine fumes.

Identify. Once the scene has been safely approached by responders and if there is a strong suspicion of a hazardous substance incident, they may alert the hazardous materials response team. However, if further information is required, the first responder can start the identification process. Text Box 7-3 outlines some of the resources available for hazardous substance identification.

Decide. The final step in the RAID process is decide. At this point, the responders should have a reasonably good idea if they are equipped and trained to proceed and take action. If the hazard exceeds the capabilities of the responders, specialized teams should be called.

Awareness of safety issues at a hazardous substance emergency is important for all responders. Individual departments must access their level of response and develop appropriate safety procedures.

TEXT BOX 7-3 SOME RESOURCES AVAILABLE FOR HAZARDOUS SUBSTANCE IDENTIFICATION.

Occupancy and Location. The mere location of an incident or the occupancy of the building or vehicle may provide clues to identification.

Placards and Labels. Hazardous substances in transit will be placarded according to U.S. Department of Transportation guidelines (see Figure 7-2). However, if the load is less than 1,000 pounds, some materials are not required to have a placard. If the placard contains the four digit identification number, the product can be identified. If not, then just a generalization as to the type of material is provided, for example, a flammable gas. Labels are similar to the placards but are found on the individual packages.

Shipping Papers. Regardless of the mode of transportation, some type of shipping papers should be with the operator; therefore in a train they are in the engine or in a truck they are the cab. These papers provide identification of the product as well as hazards.

Figure 7-2 *Placards and labels help responders identify products.*

National Fire Protection Association (NFPA) 704 System Placard. Fixed facilities may have an NFPA 704 placard on the area where hazardous substances are stored or used. This placard is a diamond divided into four small diamonds. Each diamond is color coded for the hazard and assigned a number from 0 to 4, with 0 being no hazard and 4 being an extreme hazard. The color code is red for flammability, blue for health, yellow for reactivity, and the white diamond for special information.

Material Safety Data Sheets. Fixed locations storing or using hazardous substances are required to maintain material safety data sheets, which provide product identification information, hazards, and procedures for emergencies.

Employees and Occupants. Employees working at the incident location may also be able to provide product identification information. Remember, however, that employees often downplay the event. Information should be backed up with reference information. Certain facilities may also have internal fire brigades.

ChemTrec. 1-800-424-9300 may be called for assistance in chemical emergencies not only to assist in identification but also for emergency action. ChemTrec can usually put the caller in touch with product experts or shippers.

Reference Books. Although a number of reference books exist, no emergency response unit should be without, at a minimum, the Department of Transportation's *Hazardous Materials Guidebook,* the orange book.

TECHNICAL RESCUES

Technical rescues include rescue from structural collapse, trench rescue, high- or low-angle rescue, confined space rescue, rescue from machinery, wilderness rescue, and rescue from water. The NFPA's standard *Operations and Training for Technical Rescue Incidents* (NFPA 1670), addresses technical rescue operations. Each of these types of incidents requires a great deal of expertise and specialized equipment to handle. These incidents present a number of safety issues to the responder. The safety program manager, along with the organization's management, must determine the level of response to be provided and equip and train personnel accordingly (see Figure 7-3).

■ Note
Rehabilitation is of prime concern during these incidents.

Each technical rescue response presents different safety issues from air supply in confined space to the threat of secondary collapse in structural collapse rescue. Generally, technical rescue incidents last longer than fire and emergency medical alarms. Rehabilitation is of prime concern during these incidents. Technical rescue incidents must be managed with an incident management system (IMS) and a safety officer in place. The safety officer should be familiar with the

Figure 7-3 *Technical rescue operations require specialized training and equipment.*

operation in question. General safety issues relating to technical rescue incidents include:

- Environmental conditions. Responders will be exposed for long periods of time.
- Stability of the building, trench, or confined space.
- Duration of available air supply, if necessary.
- Resources of personnel and equipment.
- Safe atmosphere—flammable gases? oxygen deficient?
- Rehabilitation of responders?
- Backup teams similarly trained and equipped.
- Logistics, food, rest, and so forth for long-term operations.

Each technical rescue incident must be evaluated for risk versus gain. Often these incidents are actually body recovery situations and the responders' safety must be given full consideration. Technical rescue is a relatively new discipline for the emergency responder and is evolving. Suggestions for the incident safety officer assigned to a technical rescue incident are provided in Chapter 9.

HELICOPTER OPERATIONS

Emergency responders are often called upon to perform operations dealing with helicopters. These operations may include working with the helicopter during a rescue, deploying responders from the helicopter into water, setting up a landing zone for a medivac helicopter, or providing ground support for a helicopter fighting a wildfire. Clearly, some of the incidents are specialized and require a great deal of training and coordination with the helicopter crew. Some general safety precautions, however, can be taken when operating around helicopters. Primarily this section deals with ground support operations and approaching the aircraft. The most common ground support operation involving helicopters is the establishment of landing zones. Several safety principles present during this function.

Landing Zone

Be sure to select a proper landing zone. This zone will differ with jurisdiction and between day and night operations. Generally speaking, the size of the landing zone depends on the type and size of the aircraft. Landing zones can range from 60 to 120 feet square. The landing zone should be on as level ground as possible. A pitch of greater than 8° to 10° is risky. The landing zone should be free from flying debris. Remember, if the aircraft weighs 2,000 pounds, then 2,000 pounds of air must be displaced for the aircraft to fly. This produces very high wind on landing and takeoff. Garbage can lids and other debris can be picked up and moved about in this wind, possibly even into the blades.

■ Note

The landing zone should be free from flying debris. Remember, if the aircraft weighs 2,000 pounds, then 2,000 pounds of air must be displaced for the aircraft to fly.

Marking the landing zone is based on local procedures. Many jurisdictions require five markers, one on each corner of the landing area and the fifth to be placed on the windward leg in the center to show wind direction. At night, the landing zone markers must be lit. Usually the marker lights are blue. Some jurisdictions mark the landing zone with headlights of vehicles. Although this practice is acceptable, it must be cautioned that the lights must stay so aimed as to not shine up into the pilot's eyes. It is not good practice to use flares or cones.

Fire suppression must also be available at the landing zone. The fire suppression unit should be specifically assigned to landing zone duties and an IMS should be in place. The crew should have full protective clothing on and be prepared to deploy a hose line for fire suppression. The landing zone fire suppression unit should be so parked as to not be a part of an emergency should one occur. Responders standing by or working near the helicopter should be provided with eye and hearing protection in addition to firefighting PPE. Figure 7-4 shows a typical landing zone configuration for a night operation.

■ Note

Responders standing by or working near the helicopter should be provided with eye and hearing protection in addition to firefighting PPE.

Crowd Control

The landing of a helicopter often draws many onlookers. Vehicle and crowd control are an important function around the landing zone. Often a helicopter will land and keep the motors running so there is no delay in departure. Nonessential people, including bystanders and responders should be kept a safe distance from the aircraft. Although this crowd and traffic control is, in most cases, a police function, other emergency responders may find themselves involved. It is a good idea to keep bystanders and nonessential emergency personnel at least 200 feet from the aircraft.

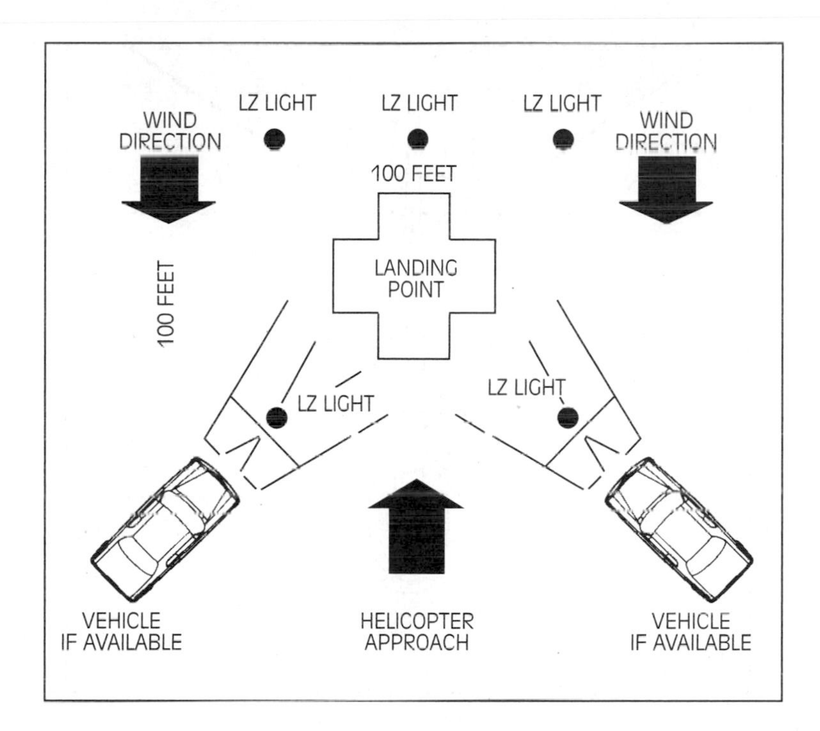

Figure 7-4 *Typical landing zone configurations for night operations.*

Approaching the Aircraft

In many cases emergency responders will, at some point, have to approach the aircraft, whether it be to load a patient or hook up a hose line to refill a drop tank. The safety principle to be applied in this case is to NEVER approach the aircraft without permission from the pilot. After receiving permission from the pilot, the responder may approach the aircraft, but should always approach from within the pilot's view and always from the downhill side if on a grade. Figure 7-5 depicts safe and unsafe approach areas to helicopters during ground operations. Should a rescue basket be used, static electricity is a safety concern. The basket should touch the ground before a rescuer on the ground touches it.

CIVIL DISTURBANCES

civil disturbances uprisings of civilians that often lead to hostile acts against law enforcement and emergency responders

There have been a number of **civil disturbances** in the United States over the years. Emergency service responders are often on the front lines during these events, either for the treatment of injured or for fire suppression. These incidents require a great deal of coordination and cooperation with law enforcement.

Emergency response personnel operating during civil unrest are exposed to many hazards, including gunfire, being assaulted with objects that have been thrown, and increased hazards at fires due to the use of accelerant. Personnel may minimize some of these risk through prioritization of incidents, police protection, and additional PPE.

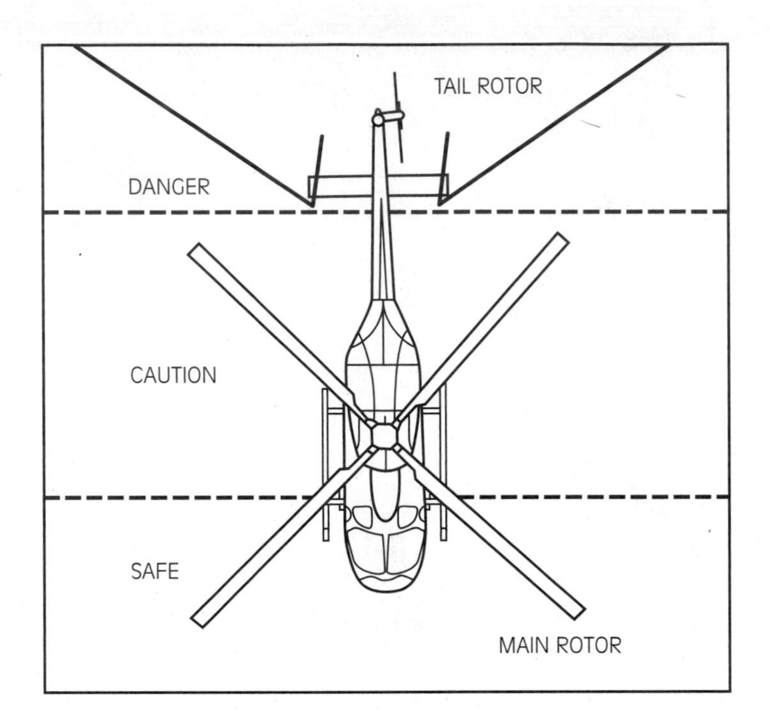

Figure 7-5 *Approach helicopters only from the safe zone.*

Incidents should be prioritized according to the potential risks involved. If a life threat is not present, the risk to personnel should be closely examined before the responders become the life safety concern. If the incident is a priority incident and units must respond, the police should either escort the units or ensure the area is clear prior to fire or emergency medical services (EMS) response. Emergency response units may be put into task forces in order to respond in higher numbers. This arrangement might be two engines, a ladder, a medic unit, and a battalion chief, with police escort. In the case of priority building fires, units should not commit to long operations, but instead hit fires with booster tanks and master streams. This will allow for a fast retreat, should the situation dictate.

■ Note
Police protection is a critical component of the operational plan at civil disturbances. This protection should be planned before an event.

Police protection is a critical component of the operational plan at civil disturbances. This protection should be planned before an event. The police often have intelligence as to where problem areas may be and what might be expected. This information should be communicated to the fire and EMS agencies. Communications and a unified command system will ensure that responders are working under the same plan and all agencies have the same information.

Some agencies that do not normally wear body armor may elect to issue the body armor for use during civil unrest. Although self-contained breathing apparatus (SCBA) would provide the necessary protection, emergency responders may wish to wear masks to limit their exposure to tear gas. These masks are lighter than SCBA and provide protection from the irritation of tear gas.

Figure 7-6 *Terrorist acts are designed to create media attention and require large commitments of resources. This photograph was taken outside the World Trade Center in New York City after the bombing.*

TERRORISM EVENTS

The explosions that shook the World Trade Center and the federal building in Oklahoma City made the nation and the nation's emergency responders aware that terrorism can occur anywhere. Terrorist acts can be classified into three categories, depending upon the material of mass destruction used. Generally, terrorist acts are either nuclear, biological, or chemical (NBC). Acts of terrorism are generally committed to attract attention, and therefore are designed to hurt or kill large numbers of people (see Figure 7-6). In an Atlanta bombing, a secondary device detonated after emergency responders were at the scene. Whether the secondary device was intended for responders is unknown, but the potential does exist that it was.

Emergency responders must, at a minimum, be provided with awareness training for terrorism. Many NBC events require the same safety precautions that one would take when responding to a hazardous materials incident, which most terrorism events are. Responders must remember that the terrorism act is designed to create a media event, and emergency responders can easily be a target. A number of resources are available to the safety and health program manager for training in safe operation at these events, including training from the United State Fire Administration and the text *Preparing for Terrorism* by George Buck.[1] As first responders, the emergency service will be the first agency on the scene of a terrorist act. We must be prepared.

---

[1] Buck, George. *Preparing for Terrorism.* Albany, NY: Delmar Publishers, 1997.

Summary

For the purpose of this chapter, response to specialized incidents includes responses to hazardous materials incidents, technical rescues, helicopter operations, and civil disturbances. Each of these requires specialized response and produces numerous unique safety concerns for the responders. Some of these types of incidents are not handled by every response agency. In some agencies that do respond to these incidents, the response may be with a specialized team. It would be hard to imagine any response agency that does not have the potential to respond to hazardous materials emergencies. The key to safe handling of these incidents is through Recognition, Approach, Identification, and a Decision as to whether on-scene resources can deal with the situation. Hazardous materials operations are governed by federal regulations that require responders to be trained to a minimum level, operate with proper PPE and in a buddy system, and have site-specific safety plans and an IMS system. Training is based on five levels of response. The safety program manager must access the responders level of training and ensure it is commensurate with the type of operations being performed.

Technical rescue is a specialized area of response that requires specialized training and equipment. Technical rescue includes rescue from a collapsed building, trench rescue, confined space rescue, high- and low-angle rescue, rescue from entrapment in small machinery, wilderness search and rescue, and water rescue. Each of these types of technical rescue are disciplines within themselves. From a safety standpoint, although some generalization regarding safety can be made, technical rescues require a great deal of expertise on behalf of the responder, the incident commander, and the assigned incident safety officer.

Many departments are using helicopters for fire suppression, and, more likely, for medivac operations. At a minimum, emergency responders may be required to provide ground support for these operations. Critical ground support safety issues relate to the selection and protection of a landing zone, crowd control, and the approach to the aircraft. Local conditions and the operator of the helicopter dictate specifics regarding these issues.

The final subject in the special operations chapter is that of response and operations in areas of civil disturbances. These situation require close coordination with law enforcement, including unified command in the IMS. Through the use of prioritization of calls in the affected area, the use of alternative tactics, and increased PPE, a level of safety can be provided.

Terrorist acts also require close coordination between law enforcement and the emergency responders. Acts of terrorism may include nuclear, biological, or chemical weapons of mass destruction. Intended to create media attention, these incidents often involve mass casualties and may have secondary devices designed to injure emergency responders.

Concluding Thought: Emergency responders are required to handle much more than fire and EMS incidents. The issues related to safety increase with the various other incidents to which we must respond.

Review Questions

1. A person who normally responds to releases of hazardous substances and takes offensive action by plugging a leak should be trained to what level?
   A. First responder operations
   B. First responder awareness
   C. Hazardous materials technician
   D. Incident commander
2. A person who normally responds to releases of hazardous substances and takes no offensive action other than notification to start the response plan should be trained to what level?
   A. First responder operations
   B. First responder awareness
   C. Hazardous materials technician
   D. Incident commander
3. What does the acronym RAID stand for?
4. List and explain three pieces of information available to help the responder identify a released hazardous substance.
5. In terms of terrorism, what does the acronym NBC stand for?
6. What seven areas of rescue are covered by the term technical rescue?
7. In general, what is the minimum space requirement for a helicopter landing zone?
8. What action must be taken before approaching a helicopter with a rescue basket deployment?
9. What PPE should be worn during helicopter landing zone operations?
10. List three hazards present during a civil disturbance.

Activities

1. Review your department's standard operating procedures (SOPs) regarding the response to hazardous materials incidents and compare the SOP to what really happens in the field. Do the responders comply? Are they trained to the level required for the response?
2. Access your department's response levels to technical rescue incidents. What services or types of response are provided? What safety issues have been identified and what procedures are in place to reduce the hazards?
3. Review your department's procedures for helicopter operations. Are there any deficiencies based on the information in the text?
4. Access the potential for civil disorders and terrorism events in your community. What procedures are in place to ensure a safe response?

Postincident Safety Management

Learning Objectives

Upon completion of this chapter, you should be able to:

- List the safety and health considerations when terminating an incident.
- Describe the demobilization process.
- Compare the concept of first in-last out with first in-first out.
- Explain the need and the process used for postincident analysis.
- List the components of a postincident analysis.
- Describe the advantages of a critical incident stress management program.
- List the key components in a critical incident stress management program.

CASE REVIEW

On Saturday June 17, 1972, at 2:35 P.M., the Boston Fire Department transmitted Box 1571 for a fire at the Hotel Vendome, Commonwealth Avenue at Dartmouth Street. It would be the first of four alarms required to extinguish a raging fire at the hotel. It took nearly 3 hours to control the fire. Sixteen engine companies, five ladder companies, two aerial towers and one heavy rescue responded by the time the fourth alarm was struck. Once the fire was out, the firefighters began a routine overhaul operation. At 5:28 P.M., without warning, the southeast section of the building collapsed, causing the worst loss of life at a single incident in the history of the Boston Fire Department.

The Vendome tragedy exemplifies the risk intrinsic to the firefighting profession and the accompanying courage required in the performance of duty. Nine firefighters were killed on that day, eight more were injured; eight women were widowed, twenty-five children lost their fathers; a shocked city mourned before the sympathetic eyes of the entire nation.

Although this tragic event occurred while units were still at the scene, it emphasizes the point that injuries and fatalities occur after the incident is brought under control, a time when we should operate more safely.

INTRODUCTION

The last step in the safety and health program consideration in terms of incident response is the consideration for postincident activities. Included in this chapter are suggestions for terminating the incident, postincident analysis, and critical incident stress management. Postincident safety and health considerations can be easily overlooked if not integrated into the total safety and health program. In other words, responders tend to let their guard down after the incident is over. Therefore, the safety officer and supervisors must be trained, and safety during this phase should be emphasized and committed to as part of the program.

■ Note
Responders tend to let their guard down after the incident is over.

INCIDENT TERMINATION

Incident termination can be comprised of three stages: demobilization, returning to the station, and postincident analysis. Postincident analysis is discussed in detail in the next section of this chapter.

Demobilization

demobilization
the process of returning personnel, equipment, and apparatus after an emergency has been terminated

Demobilization is the stage in the incident when the incident commander, in a large incident, with the recommendation of the planning section, evaluates the on-scene resources and compares them with the current situation (see Figure 8-1). Using a fire scenario, demobilization usually occurs when the fire has been brought under control and some of the firefighter resources can be placed available and returned to stations. There are numerous safety and health issues that the

Figure 8-1 *Demobilization is the stage in the incident when units begin to leave the scene.*

incident commander should address during this stage and that should be part of the department's operating procedures and safety policies.

One consideration is the sequence in which operating crews are released. The fire service has historically used the **first in–last out** approach. Very often, the unit that is first on the scene is the last unit to leave. Lately some departments have begun to change this approach, using first in-first out as a demobilization tool. This process is good because it permits the units that arrived first, and probably have worked the longest and generally under the worst conditions, to be relieved first. Some arguments are made regarding this process as the first arriving apparatus are normally committed to firefighting and have ladders raised to the building or hose connected. This argument can be overcome when departments have **standardized apparatus** and equipment. In this case it should make no difference that a first-arriving crew could later leave in another crew's apparatus. The first in–first out approach is a very important consideration during periods of extreme heat or cold and high humidity.

first in-last out
the usual method of crew release used at an emergency scene; basically, the first arriving crews are generally the last to leave the scene

■ Note
Lately some departments have begun to change this approach, using first in–first out as a demobilization tool.

The incident commander and the planning officer should also consult with the on-scene safety officer during the demobilization process. The safety officer may be more aware of crews that need to be relieved or that have operated for extended periods. The safety officer can obtain some information from the rehabilitation officer who should have been monitoring the status of members of each crew.

The use of rapid intervention crews or teams (RIC) was discussed in Chapter 5. During demobilization, the incident commander should make an informed decision as to when the RIC is no longer needed. It is not uncommon for injuries to occur long after the initial fire fight is over. In fact, some disastrous collapses

standardized apparatus apparatus that has exactly the same operation and layout of other similar apparatus in a department; For example, all of the department's pumpers would be laid out the same, operate the same, and have the same equipment; useful for the situations when crews must use another crew's apparatus

in which crews have been trapped and killed have occurred during the overhaul phase of a fire. One such incident is noted in the case review of this chapter.

Once released from the emergency scene, or in the case of a medical incident once the call has ended, there are safety concerns as well. The supervisor or senior person assigned to the crew must ensure the readiness of the team for the next incident. After any incident, the crew members should be questioned about injuries they may have incurred or any feelings about the incident that may prompt the supervisor to start the critical incident stress management process. In the case of an incident where crew members have the potential to be exposed to infectious diseases, the crew should do as much personal decontamination as possible at the scene or at the hospital prior to returning to the vehicle. At this point the supervisor or company officer must make a decision as to whether the unit can be placed in an available status or if crew members or equipment are not ready. It is not uncommon, in the emergency service field, to check equipment and stock after a call to ensure readiness. The human safety elements cannot be overlooked in this process.

Note

The supervisor or senior person assigned to the crew must ensure the readiness of the team for the next incident.

Returning to Station

The supervisor should also be alert to signs of **critical incident stress.** Vehicle operators must maintain an alertness to the road (see Figure 8-2). Reviewing the NFPA injury and death statistics, the category for vehicle crashes is titled "responding/returning." Injuries do occur during the returning-to-station stage of the incident. Returning to the station often provides the first chance for the crew to discuss the incident among themselves.

critical incident stress stress associated with critical incidents, such as the injury or death of a coworker or a child

Figure 8-2 *Safety is also a consideration during the trip back to the station and putting equipment back on the apparatus.*

Once back at the station, the apparatus is made ready for the next emergency, supplies are replaced, and equipment cleaned. Again, the safety program must incorporate the human needs at this point. The supervisor should be alert for signs of fatigue and the crews should be assessed again for any injuries or illness that may be attributed to the incident. Remember, emergency service responders often do not complain because they see minor injuries as part of the job. The supervisor should be alert to visual signs such as limping or favoring an arm or leg, cuts, bruises, and general differences in the responder. Nine out of ten times the responders are just tired, but sometimes an injury can be uncovered and reported simply by asking the right question.

postincident analysis a critical review of the incident after it occurs; the postincident analysis should focus on improving operational effectiveness and safety

Once the equipment and crews have been taken care of, the unit is readied for the next assignment. Before long however, the supervisor should sit down and discuss the incident with the crew in an informal manner. This is the first step in the **postincident analysis.**

POSTINCIDENT ANALYSIS

A postincident analysis should be done on any incident to which multiple units have responded and operated at, and any incident involving a serious injury or death. These incidents may include fires, technical rescues, hazmat incidents, multiple casualty incidents, vehicle accidents, and victim entrapments. Often called a critical review, or critique, the postincident analysis (PIA) is a step-by-step look at what happened at the incident, what went right and what went wrong. The components that the PIA should focus on are:

- Resources
- Procedures
- Equipment
- Improving operational effectiveness

The term *critique* is sometimes associated with criticize, therefore the term *postincident analysis* is used in this text.

There are two steps of the PIA. The first is the informal discussion among the crew after the incident (see Figure 8-3). Although informal, this is a good time to review what each member saw and did and what the unit did as a whole. The supervisor during this discussion should make notes for use at the formal PIA. This process can, and should, also occur in volunteer organizations.

Usually within some fixed time after the incident, all crews that were present at the incident should get together for the formal PIA. This PIA must be attended by all those who carried out a specific task at or for the incident, including those members who filled positions within the incident management system (see Figure 8-4). Often the incident commander will call for and run the PIA. It is better in most cases to have an objective person other than the incident commander facilitate the PIA as this then permits a free exchange of information and

Figure 8-3 *The informal PIA may take place at the kitchen table after the incident.*

Figure 8-4 *The formal PIA should involve all members that operated at the incident.*

allows the incident commander to be more involved in the PIA as opposed to having to lead the analysis. The PIA events should be recorded for future changes in procedure or policies or to reaffirm current ones. Some departments videotape the PIA and then distribute the tape departmentwide so that every member can learn from the incident. At minimum there should be a written form that is filled out and filed for each PIA conducted.

A good procedure for the PIA is to have all the responders discuss, usually using a graphic or plot plan of the incident, what they did on arrival, what worked, and what did not work. If injuries or deaths occurred as a result of the incident, surrounding events should also be noted and be investigated. Someone present at the PIA should have a copy of the standard operating procedures

(SOPs) available and compare the actions taken to those required by the procedures. For example, if the SOP says that the first-arriving ambulance will set up a treatment area, and in the case in question, the first-arriving ambulance entered the building to fight the fire, then either the SOP needs to be reevaluated or the crew should state the reasons for their actions.

The objective of the PIA should be to highlight positive and negative outcomes of an incident. The PIA can also provide a tool for future training and show a need for changes in procedures and policies or even a need for a new piece of equipment. The PIA should be a positive process, allowing all attendees to have a chance to speak and discuss concerns, particularly those related to short- and long-term safety and health issues. The PIA can also aid in the investigation process into any incident injuries and fatalities and become cause to develop procedures that may prevent them in the future. See Figure 8-5 for the relationship between SOPs, training, operations, and the PIA.

CRITICAL INCIDENT STRESS MANAGEMENT

stress
the body's reaction to an event; not all stress is bad, in fact some level of stress is necessary to get a person to perform, for example, the stress associated with a report that is due is often the motivating factor in doing it

Stress is a normal part of everyday life. However too high a stress level leads to negative reactions, such as actual physical illness, job burnout, and lack of productivity. In terms of safety and health, a stress management program is necessary, including provisions for employee assistance programs as discussed in Chapter 4. This section of the text focuses on after-incident critical incident stress management and the associated health and safety considerations.

Emergency responders are expected to tolerate certain levels of stress. However, some events that are usually high in stress levels are those that are powerful in terms of emotion or a combination of many smaller events. These events are called *critical incidents*. Many years ago, responders were expected to handle these events on their own and were often criticized if they displayed weakness or emotion toward these critical events. Fortunately, this is no longer the case and critical incident stress management has become not only accepted in the emergency service arena, but is a part of most organizations.

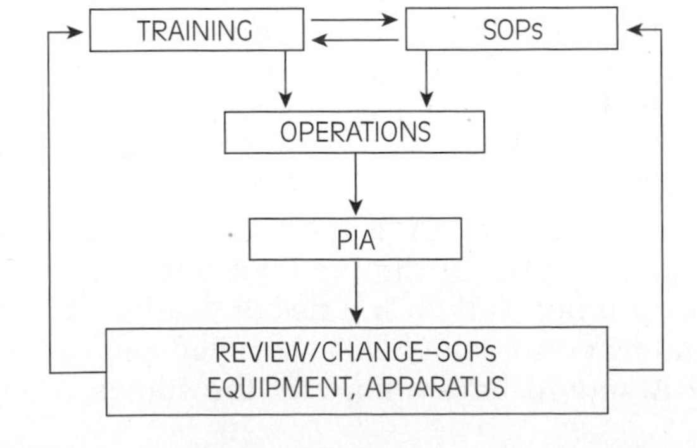

Figure 8-5 *The relationship between the PIA, SOPs, training, and operations.*

Events that typically result in critical incident stress include:

- Traumatic death or severe injury to a coworker, particularly those suffered in the line of duty (see Figure 8-6)
- Suicide of a coworker
- Traumatic death or serious injury to children
- Mass casualty events
- Prolonged events
- Death or injury to a person caused by the emergency responder, for example a traffic accident while responding to an incident
- Events in which there is a great deal of media attention
- Personally significant events, for example the death of an elderly man when the responder just lost an elderly family member

When a critical incident occurs it is important that a process be in place for both stress defusing and debriefing. This process should include access to the Critical Incident Stress Management (CISM) system. In some cases, this debriefing should take place before the responders have to ask for it, as the responders may not feel they need it or may not ask for fear of showing signs of weakness. Each organization sets procedures for when and how to access the system. Some systems automatically set the process in motion when any of the critical incidents listed takes place, or at the request of an incident commander or company supervisor for any event.

Figure 8-6
Accidents involving coworkers are particularly stressful. CISM should always follow these incidents.

Without CISM, the responder may be more likely to develop negative physical, behavioral, and psychological reactions. Unwanted physical reactions include fatigue, sleeplessness, changes in eating habits, and body aches. Behavioral changes may also occur, including changes in activity levels, difficulty in concentration, nightmares, flashbacks, memory problems, and isolation. Psychological symptoms include fear, guilt, sensitivity, depression, and anger. In some extreme cases, if not properly managed, a person suffering from critical incident stress reactions may begin to abuse alcohol and drugs and may resign from employment in the emergency service field. The safety and health program that includes CISM helps the responder deal with the issues associated with the event and hopefully avoid many of the unwanted reactions.

peer defusing
the concept of using a trained person from the same discipline to talk to an emergency responder after a critical incident as a means to allow the responder to talk about his feelings regarding the event in a nonthreatening environment

There are different types of CISM both at the scene and after the event. The first should occur at the scene and is termed **peer defusing.** Peer defusing involves an informal discussion about the event by a group of peers who experience the event together. This phase often occurs normally as the incident deescalates and responders get together and discuss the event. The second is the formal debriefing that usually occurs sometime after the incident, maybe even a few days, and is the formal process in which only the responders to the incident get together with a peer debriefer and discuss their roles at the incident and the overall response. Two issues are very important in this phase: (1) All those involved should try to attend, and (2) the peer counselor should have had training in this type of operation.

Note
All those involved should try to attend, and the peer counselor should have had training in this type of operation.

Very often CISM teams are regionalized, which allows for a peer debriefer from the same occupation but an outside agency to serve in that capacity. It would be difficult for a department to have an inside team, as each debriefing would involve a peer from the same department who was not at the incident. This approach is not good. Good CISM team characteristics include regionalization, a mental health professional as a team member, nonpartisan, and having received appropriate training in CISM. Total confidentiality and relative anonymity are vital elements of a debriefing and the peer debriefer must follow and buy into these concepts, as should the management of the response organizations.

Note
Total confidentiality and relative anonymity are vital elements of a debriefing and the peer debriefer must follow and buy into these concepts, as should the management of the response organizations.

A third type of CISM is on-site defusing, which can be useful at large, long duration events. As they leave the operational area, responders are evaluated by a member of the local CISM team for signs and symptoms of distress and treated appropriately. The fourth type of CISM is demobilization. Demobilization allows responders after operating at a large-scale event to have a buffer period of 30 to 40 minutes before leaving the location. This period can provide a time for nourishment and brief education on the signs and symptoms of critical incident stress.

CISM is a relatively new concept in emergency response. Although the incidents have occurred forever, it was not until the 1980s that incident stress was recognized as a problem. The safety and health of the employees and their longevity in the response system depend on CISM. It is a necessary component to any safety and health program.

Summary

Safety and health after the incident are sometimes overlooked. During this phase of the response there is potential for injuries to occur and, therefore, this must be considered in the safety and health SOPs. Postincident safety considerations begin with the termination of the incident, at which time the incident commander makes a decision about the demobilization of resources, and these resources return to their stations. On a single unit medical type incident, the incident is terminated once the patient has been handed over to the emergency department or to a transport provider. It is important during this phase of the operation that the same consideration be given to the human components of the crew that is given to the equipment and supplies.

Part of the postincident safety considerations is postincident analysis (PIA) and critical incident stress management (CISM). The PIA can begin when the unit returns to the station and the superiors have an informal discussion with the crew regarding the incident. Later, a formal PIA must be conducted using a prescribed format and documented for future use in training or for updating SOPS. All personnel who responded to the particular incident must be included in the formal PIA.

Another safety and health concern postincident is critical incident stress management. Critical incident stress management programs are integral to the safety and health program. Teams should be available to assist members following incidents that generate high levels of stress. Often these teams are formed on a regional basis to allow for peer reviewers from outside of one's own agency. Supervisors must have training in and be aware of common incidents that can lead to critical incident stress reactions. All personnel must be aware of available CISM programs and how to access them at any time, given the need.

Concluding Thought: Many essential safety and health program functions occur after the incident itself, but are just as important as those that occur before or during an incident.

Review Questions

1. What are the three stages in terminating the incident?
2. Differentiate between the first in–first out and first in–last out approach. List advantages and disadvantages of each.
3. It is best to allow the incident commander to facilitate a PIA.

   A. True

   B. False

4. Using a supervisor from within the organization as a peer debriefer is an important step in the CISM process.
   A. True
   B. False
5. List three qualities of a good CISM team.
6. The RIC can be released once a fire is declared under control.
   A. True
   B. False
7. Describe the relationship between training, SOPS, operations, and the PIA.
8. Inasmuch as possible, decontamination should occur before leaving the scene.
   A. True
   B. False
9. Who should be present at a formal PIA?
10. The informal PIA should occur upon return to the station before the unit is placed back in service.
    A. True
    B. False

Activities

1. Review your department's SOP for demobilization at emergency incidents. Which process is used, first in–first out or first in–last out? Consider changes. Is the approach used in your department working? What roadblocks might prevent a first in–first out procedure?
2. Develop a postincident analysis SOP for your department. If there already is one, how effective is it? Are the objectives noted in this text being achieved?
3. Research information on CISM teams in your area. Have personnel had training in recognizing critical incident stress reactions and how to access the team?

Personnel Roles and Responsibilities

Learning Objectives

Upon completion of this chapter, you should be able to:

- List the roles and responsibilities of the individual responders and their relationship to the overall safety and health program.
- List the roles and responsibilities of the supervisors and their relationship to the overall safety and health program.
- List the roles and responsibilities of the emergency service management and their relationship to the overall safety and health program.
- List the roles and responsibilities of the incident commander and his or her relationship to the overall safety and health program.
- List the roles and responsibilities of the safety program manager and his or her relationship to the overall safety and health program.
- List the roles and responsibilities of incident safety officers and their relationship to the overall safety and health program.
- List the roles and responsibilities of the safety committee and its relationship to the overall safety and health program.

CASE REVIEW

The department had everything in place. The board of fire commissioners had decided several years earlier that the department had grown to a size that the position of training/safety officer was needed. The department did a promotional process and the assignment was made.

The training officer immediately set out to improve the department's training and safety program. In terms of a safety program, the training/safety officer started a safety committee, performed a safety audit on the department based on the National Fire Protection Association's (NFPA) standard 1500, and introduced the concept of risk management. After laying this basic groundwork, the officer, with the aid of the safety committee, developed a number of very good programs.

The training/safety officer made a mistake during the implementation of the programs. Specifically, the training/safety officer failed to identify the roles of the various employees in the department. Some of the feedback he was getting was that, although the statistics for the department were getting better, many of the operations people on a particular shift were not following some of the basic safety procedures and policies, such as using seatbelts and using personal protective equipment (PPE) on medical alarms. The safety officer was concerned that if the procedures were not enforced on some responders, other members would follow the procedures only when convenient.

But who was at fault? The safety officer began at the first-line supervisor level and, with training, reemphasized the importance of the program and the importance of enforcement. The training sessions were well received, but the problem continued. The safety officer lacked supervisory authority over the first-line supervisors, so he went to his boss, the director. As he approached his meeting with the director, he realized that the director had never been invited to any of the safety training sessions, and had not had any input on any of the safety policies. In fact, he did not know the director's position on safety at all. The safety officer explained to the director that the first-line supervisors were not following proper procedures regarding safety programs, and he felt that it was time to start a formal discipline process. The director disagreed. In fact, he told him that many of the safety procedures that had been developed were not necessary, that none of these programs were needed, and that getting hurt was part of the job.

The safety officer left in amazement. How could this be? The director promoted a training and safety officer and now thinks that the safety program is not needed?

What the safety officer did not understand was that the director did not ask for a safety officer, the board of fire commissioners did. In fact, the director often said in the station, "This safety business will be the ruination of the fire service." Many members heard the director's remarks and many agreed with him. Hence, many supervisors and members, knowing the director's position, did not follow the procedures.

Note
Safety and health responsibility extends from the new individual member to the top of the organization, whether that top is the chief or other titled director.

INTRODUCTION

As part of putting a safety program together, the safety program manager must have an understanding of the various roles played by the members of the organization (see Figure 9-1). Who exactly is responsible for the safety and health program within an organization? After reviewing the information that has been presented thus far, the answer should be clear: everyone. Safety and health

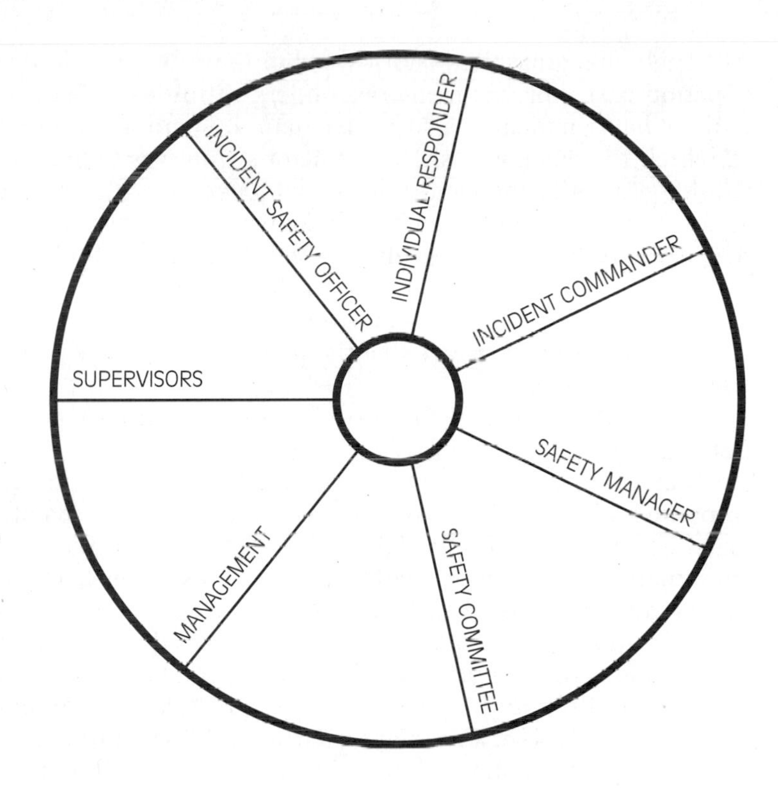

Figure 9-1 *The interrelationship of roles can be likened to a spoked wheel. The wheel will only be as strong as the weakest spoke.*

■ **Note**

It is not important how many policies are written, how many safety officers are assigned, or how much money is put into a program; without compliance and a safety attitude on behalf of the responders, the program will not be effective.

responsibility extends from the new individual member to the top of the organization, whether that top is the chief or other titled director. The interplay of the various groups and individuals in the safety and health program can be compared to a spoked wheel: The wheel is only as strong as its weakest spoke.

Although there are differing responsibilities in the various organizational levels, the goal of the program remains the same. This chapter examines these various levels and the associated roles that the persons in them may play.

INDIVIDUAL RESPONDERS

Individual responders are probably the most important link in the safety and health program. This may sound unusual, however, the individual responders can have a great impact on the entire program. It is not important how many policies are written, how many safety officers are assigned, or how much money is put into a program, without compliance and a safety attitude on behalf of the responders, the program will not be effective.

The SAFEOPS approach was introduced in Chapter 1. The A in this mnemonic device represented attitude. Attitude may well be the most important obstacle to overcome when dealing with safety and the individual responder.

Although changing, the historical belief that injuries and death go with the occupation is still part of many responders' attitudes. The belief or attitude that it will not happen to me is also human nature. There was once a line in a war movie in which the actor was shot. The actor's line was "I'm shot. I never thought I would get shot!" One would think that a person fighting a battle in a war could assume that the potential for being shot was reasonably high, yet he was surprised that it really happened to him. We see this same belief with people, whether involving crime, car accidents, or, in some cases, being injured at work. In order for a responder to want to operate safely and comply with procedures and polices, the responders must realize that the potential for injury to him or her is real and could happen.

Individual responders must also realize that they are members of a team. Depending on the type of response unit, this may be a two-person team on an ambulance or a five- or six-person team on a fire unit. Although an individual, the responder's action almost always has some impact on the other members of the team. For example, if a firefighter does not wear all of the necessary protective equipment and is overcome with smoke, the rest of the team has to rescue the firefighter, and the entire team is affected.

The individual responder should also be willing to take an active role in the safety and health program. This may involve making suggestions, working on the safety committee, and using good safety practices. It is important that the more senior responders set an example, particularly when training a new recruit. Text Box 9-1 describes some individual responder dos and don'ts.

TEXT BOX 9-1 DOS AND DON'TS FOR INDIVIDUAL RESPONDERS.

Do Be:

- *An active team player.* Be prepared to fill your role and watch out for team members, including following safety-related procedures. If a team member does not have protective gloves on when starting an intravenous line (IV), remind her.
- *A good communicator.* This applies to both listening and speaking. When working as a team, communicate with other team members. For example, when raising a ladder or lifting a stretcher, maintain contact with team members and ensure that the team makes the lift together on command.
- *A leader.* Display safety leadership. Set the example.
- *Aware of surroundings.* Maintain a constant awareness of surroundings for both yourself and the team. Look for danger signals, communicate concerns.

- *A responder who works within individual abilities.* If you know you cannot do something alone, get help. Countless back injuries associated with lifting could have been avoided when others were available to help.
- *Together.* Stay with your assigned team, no freelancing.
- *A responder with a Safety Attitude!*

Don't Be:

- *Preoccupied.* Emergency responders are called upon at a moment's notice. Be thinking about the incident and the potential outcome. Put everything else out of your mind.
- *Complacent.* Do not let the routine nature of a medical incident allow you to let your guard down. Do not assume that the fourth fire alarm today in the same building is not a fire.
- *Surprised.* Many events that occur at emergency incidents are predictable—a disorderly crowd in a bar, a building collapse after exposure to fire. Predict and be prepared.
- *A responder who attempts to do things outside of personal abilities.* If you cannot swim, do not jump in the water to rescue someone.
- *Maintain an attitude that it will not (cannot) happen to me.* Maintain a Safety Attitude.

SUPERVISORS

Although many members of the organization fall into the category of supervision, for the purpose of this chapter, the term supervisor is used to describe first-level or frontline supervision in the emergency response organization. For fire service departments this may be a company officer, lieutenant, or captain assigned to a firefighting unit. In the case of sole emergency medical services (EMS) providers, it may be a senior paramedic or crew chief. Regardless of title, the safety and health roles and responsibilities for the first-line supervisor remain the same (see Figure 9-2).

Note
The first-line supervisor is the grassroots person who supervises the team.

The first-line supervisor is the grassroots person who supervises the team. The supervisor ensures that team members stay together, maintains accountability, sees that seatbelts are fastened prior to apparatus moving, and ensures that personal protective clothing for the job at hand is worn. The supervisor may also be responsible for inspection of protective clothing and seeing that regular maintenance procedures are followed.

Figure 9-2 *A supervisor discusses operations with crew.*

■ Note
The first-line supervisor is the link between the organization's management and the individual responder.

The first-line supervisor is the link between the organization's management and the individual responder. The supervisor is responsible to see that procedures are followed and are understood by team members. The supervisor also communicates safety concerns from the team to higher management levels in the department.

The first-line supervisor generally has risen through the ranks and completed some type of performance-based promotional process. In some cases, supervisors might be assigned based solely on seniority. In either case, the first-line supervisor should have the necessary mix of experience and education to recognize dangers inherent in the occupation, whether on an emergency scene, responding, or in the station. The supervisor must apply this knowledge and take responsibility for the crew in terms of safe operations, including enforcing organization operating procedures, following laws, and maintaining awareness of team location and any dangers present. The supervisor must also follow the practices noted previously for individuals, with specific emphasis on leadership and setting the example.

EMERGENCY SERVICE MANAGEMENT

Somewhere in the organization above the first-line supervisory level is the organization's management. This might be the fire or EMS chief, or the director of emergency services, depending on the type and structure of the organization. For the purpose of this section, emergency service management is used to define the very top levels of the organization, or senior management staff. The safety program manager, regardless of rank or position should be a part of the senior management structure, but is discussed later in this chapter (in the section titled Safety Program Manager).

Senior management also has many roles in the safety and health program. While having the roles and responsibilities as individuals and as supervisors, the senior management staff often controls the financial resources and gives final approval for policy and procedure implementation.

The senior management staff must give the safety program priority in terms of support, both financial and administrative. This support must be realized by the individual responders at lower organizational levels. For example, the paramedic on the street must know that the EMS chief has a commitment to provide a safe work environment with the employees' health as a driving factor. Have you ever heard the statement, "We will do such-and-such because it is required by OSHA"? A better statement might be, "We will do such-and-such, because, after analysis, we have found it to be another measure of safety provided to our members." One way of communicating this commitment is through a written organizational safety policy. An example policy recommended by the Florida Department of Labor is shown in Text Box 9-2.

TEXT BOX 9-2 ORGANIZATION SAFETY POLICY.

Management Commitment and Involvement Safety Policy

It is the policy of our department to follow the highest safety and health standards. Safety does not merely occur by chance, but occurs as the result of commitment and participation. Our firefighters are our most important assets. Lost time from the job due to accidental injuries is costly to everyone. Our objective is a safety and health program that will eliminate or reduce the number of accidents and injuries to an absolute minimum.

The responsibility for a safe and healthful environment is shared throughout the fire department. The safety officer provides my ongoing direction and guidance in safety matters. I, in conjunction with the department officers, am responsible for training. All officers/supervisors shall execute and enforce these safety rules, policies, and procedures with the utmost regard for the safety and health of our firefighters.

Our departments's policy is that employees make suggestions and report accidents and unsafe conditions to their supervisor and know their rights not to perform unsafe work tasks, without fear of reprisal. I provide the leadership for the safety and health program's effectiveness and improvement to promote safe working conditions through employee feedback, safety committee reviews, and program funding. The safety officer responsible for the safety and health program for our organization is:

Ongoing safety program activities shall include:

- A safety committee with firefighter representation
- Safety orientation, education, and training
- Safety committee review of safety rules, policies, and procedures
- Appropriate corrective actions on accident reports and revising safety rules when appropriate

Firefighters are responsible for following all aspects of the safety and health program and for compliance with all rules and regulations. Responsibility for continuously practicing safety while performing their duties is a condition of employment. All violations of this manual shall be investigated and disciplinary action applied as warranted.

Fire Chief ______________________________ Date ____________

INCIDENT COMMANDER

Most injuries in the emergency service occur while operating at various types of incidents. The standards, and in some cases the regulations, require that incidents, regardless of type be managed with an incident management system (IMS). At the highest level of the IMS is the incident commander (IC). The fact that most injuries occur at the incident scene and that the IC is in charge of the incident scene leads to the conclusion that the IC has a great deal of responsibility in ensuring safety at the incident. The IC is charged with setting the overall incident strategy and strategic goals and assigning tactics to meet these goals (see Text Box 9-3). The IC must balance risks involved in meeting the strategic goals while maintaining a constant regard for team safety. A careful, but quick, risk benefit analysis must be performed before deciding on a strategy.

The IC must first decide on the basic strategy—offensive or defensive—to be employed. The IC should make this determination based on past experience and education. The IC must rely on information that may or may not be accurate at the time he receives it. Common information is the type of building, potential for occupants, the progression of the fire, and the resources available. For example, a vacant building with moderate fire involvement might immediately lead one to believe that a defensive attack is the strategy of choice. But if this building were in the middle of a block of closely built, occupied buildings, an offensive attack might be ordered.

Incident commanders must have a good knowledge of fire dynamics, building construction, and accepted tactics. They must also have a good level of field experience. Incident commanders can rely on past incidents to provide knowledge as to what is effective and what is ineffective. They can also learn from attending postincident analysis of other incidents and applying the lessons learned to future incidents. Although the decision-making process taught in many officer programs relies on a several-step decision-making process, in actuality, research shows that ICs use what is known as the recognition-primed decision-making (RPD) making process at emergency scenes. Basically the RPD theory says that ICs do not have time at an emergency to go through a several-step process with selection of alternatives, but instead, generally arrive and immediately begin to take control of the incident based on past experience with similar incidents. If something worked ten other times at a bedroom fire, it will work again. For the IC to utilize this process and to provide for sound strategic plans, simulation training is necessary, so the IC can be trained to react to predicable events at the emergency scene.

TEXT BOX 9-3 STRATEGIC DIFFERENCES.

The Incident Commander must evaluate the risks of offensive and defense fire attacks

Offensive Attack Strategy. Firefighters are in close contact with the fire and exposed to all the inherent dangers, including burns, falling objects, being lost, being overcome with toxic gases, and building collapse.

Defensive Attack Strategy. Firefighters are outside the immediate area of fire danger. Most protective equipment is still required. Dangers still exist, such as building collapse, falling objects, and exposure to heat and fire products.

SAFETY PROGRAM MANAGER

The safety program manager is the focal point for safety-related activities in the organization. As stated previously, the safety program manager should be a member of the senior staff. The program manager has the overall responsibility for overseeing the program and all associated components.

The roles and responsibilities of safety program mangers are summarized in Text Box 9-4. Depending on the size of the organization, the safety program manager may have a staff of assistants, or, in a smaller organization, may fulfill these roles in addition to other duties, such as being the training officer. In either case the safety program manager must have sufficient experience and education in risk management principles, cost-benefit evaluation, and emergency service operations to be fully effective.

TEXT BOX 9-4 THE ROLE AND RESPONSIBILITIES OF THE SAFETY PROGRAM MANAGER.

- Be the organization's risk manager
- Be a member of senior staff with direct access to the chief or director
- Receive and act on recommendations from individuals and the safety committee
- Cause safety and health policies to be developed
- Be liaison with workers' compensation providers
- Be liaison with the department physician on health-related issues
- Investigate injuries or line-of-duty deaths
- Investigate all accidents
- Maintain all records relating to health and safety
- Perform facility inspection for unsafe conditions
- Evaluate procedures from a safety and health prospective
- Perform cost-benefit analysis
- Attend and participate in postincident analysis
- Have input in to departmental standard operating procedures
- Perform evaluation and analysis of data relating to safety and health
- Maintain awareness of trends in emergency service safety and health, including new standards or regulations and court decisions

■ **Note**
Depending on the size and type of the organization, the incident safety officer may be a dedicated position from within the safety division, or maybe a first-line supervisor assigned to the safety role on an incident-by-incident basis.

INCIDENT SAFETY OFFICER

Because most of the injuries that occur to emergency service workers occur at the incident scene, the assignment and function of the incident safety officer is key to the safety and health program. Depending on the size and type of the organization, the incident safety officer may be a dedicated position from within the safety division, or may be a first-line supervisor assigned to the safety role on an incident-by-incident basis. Some departments assign extra units to large incidents and have these crews function in certain safety-related roles, including incident safety officer, accountability, entry control officer, or rapid intervention companies. In some cases, the incident safety officer is in name only, for example, when the assignment is made to anyone who is free at the incident. This is not good practice, because the incident safety officer must have additional knowledge, be well experienced with the incident at hand, and have no other duty. Different incidents may require a different level of safety officer to be assigned (see

Figure 9-3 *An incident safety officer at a vehicle accident.*

Figure 9-3). For example, a competent incident safety officer at a fire incident may not be a very good safety officer at a hazardous materials emergency. The incident and the level of expertise of the individuals should dictate the assignment.

After arrival at the incident scene, the incident safety officer should evaluate the incident and what is happening. A prediction of what could or is going to happen should also be part of this evaluation. This evaluation involves a 360° walk around the scene and examination of available information, such as preincident plans. The incident safety officer should talk to occupants or owners and question them about hazardous situations relating to the property or vehicle, for example, alternative fuel use in a car, or gunpowder in a home, or the storage of explosives. In 1997, a Florida firefighter was killed by a .22 caliber rifle in a house that was on fire and subjected to the intense heat.

Depending on the type of incident, the incident safety officer should assess the operation from a safety point of view and relay findings to the incident commander. The safety officer must be given the authority to immediately stop unsafe acts that are immediately dangerous to responders. The stopping of an assignment must be relayed to the IC immediately as well. During the incident evaluation, the incident safety officer should assess the operating personnel as well, including the use of proper protective clothing, accountability, and crew intactness. Text Box 9-5 examines some incident types and the role of the incident safety officer. One good technique to remind the safety officer of what to look for at different type of incidents is to provide a clipboard and worksheet that can be completed. These should be durable enough to be used under various weather conditions. There should be a worksheet for each different type of incident that the department may be called upon to respond to. Figure 9-4 is an example of a worksheet

Fire/EMS Department
Incident Safety Officer Worksheet
Vehicle Accidents

Scene
- ☐ Traffic Controlled
- ☐ Utilities Secured
- ☐ Hazardous Materials

Vehicles
- ☐ Stabilized
- ☐ Structural Stability
- ☐ Cargo
- ☐ Fuel Type

Operations
- ☐ IMS
- ☐ Fire Supression Support
- ☐ Extrication
- ☐
- ☐
- ☐
- ☐

Personnel
- ☐ Adequate
- ☐ Proper Training
- ☐ PPE
- ☐ Teams Intact
- ☐ Within IMS
- ☐ Accountability
- ☐ Rehabilitation
- ☐

Site Sketch

Units Assigned

______
______
______
______
______
______

Safety Issues

______
______
______
______
______
______

Figure 9-4 *The incident safety officer's checklist for a vehicle accident.*

for multiunit response to a vehicle accident with entrapment. This worksheet can later be used as part of the postincident analysis.

The incident safety officer serves as a member of the IC's command staff and must be a resource for them. The IC must use the safety officer's knowledge and expertise in helping with strategic and tactical decisions.

TEXT BOX 9-5 VARIOUS INCIDENT TYPES AND THE ROLE OF THE INCIDENT SAFETY OFFICER.

Fire Incidents

- Strategy to match situation. Do all operating members know the strategy?
- Crew intactness.
- IMS in place.
- Accountability system in place.
- Proper level of protective equipment being used.
- Building or vehicle structure status.
- Is a collapse zone established?
- Rehabilitation setup.
- Physical condition of personnel.
- Proper scene lighting.
- Communicating to operating teams of unsafe locations, i.e., holes in floors, wires down, backyard swimming pools.
- Means of egress for crews.
- Risk assessment.
- Rapid intervention crews ready.
- Utilities secured.

EMS Incidents

- IMS set up if multiple units operating.
- Accountability system as needed.
- Proper protective equipment for task at hand.
- Structural condition of crashed vehicles, broken glass, sharp metal.
- Traffic hazards.
- Reduction of exposure to bloodborne products.
- Safe number of rescues; not trying to do too much with too little.
- Leaking fluid.
- Fire suppression support.
- Infection control.
- Rehabilitation.
- Physical condition of crews.

Hazardous Materials Situation

- IMS in place.
- Accountability system in place.
- Rapid intervention team available and with same protective clothing as entry team.
- Teams operating within level of training and available resources.
- Adequate protective clothing.
- Proper zones established.
- Pre- and postentry physical condition monitoring.
- Rehabilitation available.
- Decontamination process adequate.

Technical Rescues

- IMS set up.
- Accountability system in place.
- Adequate resources, both human and equipment.
- Personnel acting within level of training.
- Proper protective equipment.
- Proper equipment available—ropes, air supply, personal flotation devices.

THE SAFETY COMMITTEE

One of the most common methods to involve employees and to have a forum for employees to share concerns about the organization's safety and health program is through the use of a safety committee. The safety committee can be a resource for the safety program manager and other safety professionals in the organization. Some safety committees, as a group, review accidents and injuries and assist the program manger with efforts to reduce them. Having employee involvement on the committee helps to deliver to all employees the message that safety is a priority and gives the employees the feeling that they are involved, through representation, in formulating procedures and safety-related policies (see Figure 9-5).

If the members of an organization are represented by a labor union, often there is a contractual requirement to have a safety committee and to have em-

Figure 9-5 *The safety committee must have members from all levels of the organization.*

■ Note
The safety committee should have representation from all levels of the organization.

ployee representation on it. Even if there is no requirement, the safety committee should have representation from all levels of the organization. Do not forget to have all administrative personnel represented, including people in no-response roles, such as dispatchers or office staff.

Once formed, the safety committee should elect a chairperson who may or may not be the safety program manager. The committee should meet as often as required to do business, but no less than four times a year. The minutes of the meetings should be published in a format that allows all members of the department to read and review them. Any recommendations from the committee should be forwarded to top management for action. Depending on the organization, the safety committee chairperson may have direct access to the top manager without going through the various levels of command. Text Box 9-6 describes the Florida Department of Labor safety committee policy.

TEXT BOX 9-6 EXAMPLE OF A SAFETY COMMITTEE POLICY.

Safety Committee

This section formally establishes the safety committee, provides a clear statement of duties and responsibilities, and outlines administrative, procedural, and reporting requirements. The safety committee will meet quarterly.

Organization and Duties

The duties and responsibilities of this committee are to:

1. Actively participate in safety and health training programs and evaluate program effectiveness.
2. Review inspections to detect unsafe conditions, materials, practices, and environmental factors.
3. Review and analyze all workplace accidents for hazards, trends, and proper corrective actions and recommend updates to safety rules and procedures as needed.
4. Recommend methods and activities aimed at hazard reduction and/or elimination.
5. Review, compile, and distribute hazard, safety, and health information to the department.
6. Review suggestion or incentive programs.
7. Review new laws, regulations, and guidelines to assess impact and develop courses of action to effectively meet and comply with requirements.

Membership

Participation on the safety committee will include, as a minimum, the following assigned members from our organization:

The safety committee chairperson is ______________________.
(May be one of the positions below.)

Position	*Name of Assigned Member*
Supervisor	______________________
	______________________
Firefighter	______________________
	______________________
Safety officer	______________________

Order of Business

The order of business for the safety committee meeting will be:

1. Record of attendance
2. Approval of previous minutes
3. Unfinished business/open action items

4. Review of accidents, equipment failure, and recalls
5. Presentation of new business
6. Reports on special assignments
7. Reports of inspections
8. Suggestions, awards, recognitions, and incentives
9. Special feature (film, talk, demonstration)

RECORD KEEPING

Formal documentation shall be kept for each meeting. Confidentiality is critical in the discussion of incidents and injuries and must be addressed both in the discussion and the writing of the minutes. Copies of safety committee minutes are to be posted in all work locations and are to be forwarded to the fire chief as well as safety committee members.

Summary

Each member of the organization has a particular responsibility in the safety and health program. From the individual responder's role in maintaining teamwork and following procedures to top-level management appropriating money for safety equipment, the program will not be effective without everyone doing their part and assuming their role (see Figure 9-6).

Some specific roles that must be considered in the program are the safety program manager, the incident commander, and the incident safety officer. The safety program manager is the nucleus of all activities relating to the program and must be knowledgeable in multiple subjects dealing with safety and risk management. The incident commander also plays a vital role as the person who selects the strategy and tactics for a given situation. Commitment to safety and an understanding of risk management is an important attribute for the incident commander. The incident safety officer responds and becomes a member of the incident commander's staff. The incident safety officer performs an evaluation of the incident and the operations, and recommends operational changes to the incident commander. The incident safety officer must have the knowledge and experience

Figure 9-6 *Like teamwork at an emergency, each member of the organization has an integral role in the safety and health program.*

specific to the particular incident at hand. There are various safety-related issues that are particular to incident type.

The safety committee is an extension of the organization's membership and safety manager. The safety committee provides a forum for members to make suggestions on safety and health matters. The committee can also be used to investigate accidents and injuries and make recommendations for improving existing or implementing new procedures designed to reduce occupational injures. The safety committee should have membership and representation from all levels of the organization, and may or may not have direct access to the organization's senior manager.

Concluding Thought: Every person in the organization plays an integral role in the overall safety and health program.

Review Questions

1. According to this text the first-line supervisor would be responsible for:
   A. Ensuring the proper protective clothing is worn at a single-unit medical incident.
   B. Communicating safety concerns on the crew's behalf to the next level of management.
   C. Ensuring that safety procedures are followed while responding to an incident.
   D. All of the above are responsibilities of the first-line supervisor.
2. List three good attributes of an individual responder.
3. List three don'ts or things that an individual responder should avoid.
4. Attacking a room and contents fire from the inside of a structure would be considered an offensive fire attack.
   A. True
   B. False
5. A __________ fire attack would be one in which the responders protect exposed structures and do not enter the fire building because of fire conditions.
6. List three things that an incident safety officer might look for at a fire incident.
7. List three things that an incident safety officer might look for at an EMS incident.
8. List three things that an incident safety officer might look for at a hazmat incident.
9. List five roles of the safety program manager.
10. Describe the relationship between the incident commander and the incident safety officer.

Activities

1. Compare the activities of your department's safety committee to the recommendations in the text. What could be improved? If you do not have a safety committee, how could one be started?
2. What are the procedures in your department for assignment of an incident safety officer? What improvements can be made? How well does the incident safety officer interface with the incident commander?
3. Examine your own roles at incidents. Do you find yourself displaying any of the don't traits listed in the chapter? What can you do yourself to improve the safety of your team?

Safety Program Development and Management

Learning Objectives

Upon completion of this chapter, you should be able to:

- List the essential elements of a safety and health program.
- Describe the process required for the development of goals and objectives.
- Develop an action plan based on the goals and objectives.
- Perform a cost-benefit analysis.
- Describe the relationship of training to the safety and health program.
- Describe the process for developing standard operating procedures.

CASE REVIEW

A small midwestern fire and emergency medical services (EMS) department realized, through risk identification and analysis, that the number of back sprains and strains occurring within their combination department were well above the national average. This department was just developing a comprehensive safety and health program and was performing the risk identification and analysis during this startup period. The department had appointed a safety program manager and a safety committee.

Both the safety program manager and the safety committee recognized the issue with the back injuries and sought solutions. First, the problem, or risk, was identified, specifically a higher-than-average number of back injuries. The safety program manager and safety committee then developed a broad-based goal of reducing back injuries by 50% the first year and an additional 25% the second year and each year thereafter. There were three goals to the program: The first dealt with training and procedures, the second with placement of equipment on the apparatus, and the third with physical readiness to perform the lifting tasks.

The first objective focused on training the employees to recognize that a problem existed. This was accomplished through a training session where the local statistics were compared to national statistics with the information provided to all members. Also at this training session, a new procedure was introduced that required any nonemergency lifting of an item weighing over 50 pounds to be done with two people.

The second objective required the committee to look at each piece of equipment and where it was carried on the apparatus. The equipment locations on the apparatus were changed so that the heaviest items were placed at or below waist level of the average employee.

To achieve the third goal, an exercise physiologist was contacted, and exercises particular to strengthening back muscles were added to the daily fitness routine of the members.

After one year the program completely eliminated all back injuries in the department. And, excluding personnel cost during training, the program did not cost the department a penny.

INTRODUCTION

This text has thus far allowed the reader to form a good understanding of what is required for a comprehensive safety and health program. Having the information presented as tools, the text now must lead the reader down the path to getting started. With all this information, it is difficult to determine just where to start.

This chapter provides just that information—what the essential elements are to a successful program, the setting of goals and objectives, action planning, overcoming barriers, performing a cost-benefit analysis, relating the training program to the goals of the safety program, and developing standard operating procedures (SOPs). Although presented in this order for the purpose of the chapter, depending on your organization, the order of the process may be different.

ESSENTIAL ELEMENTS

The first step in the process of developing the safety and health program is to determine what essential elements are needed. These may vary by department type and size. In some cases, one role may be shared by more than one person, or one person may perform more than one role. Using the information presented in this text so far, we conclude that the following elements are essential to an effective program:

- Top management that is committed to the program
- A safety and health program manager
- Some type of record-keeping system for data analysis
- Incident safety officer(s)
- A training program
- Standard operating procedures
- Proper personal protective equipment that meets requirements and standards
- A safety committee
- A department physician
- Access to local, state, and national injury and death statistics

Safety is not cheap. Some of the elements, such as a department physician can be expensive, whereas getting the commitment of top management is free. However, the program can be developed over a period of time as funding becomes available. It is better to get something started and then seek out the funding for the more expensive components. At minimum, we need the commitment of top management, a safety program manager, a safety committee, a record-keeping system, and access to data in order to get the program off the ground.

SETTING GOALS AND OBJECTIVES OF THE PROGRAM

Once the essential elements of the safety program are in place and the risk identification process is completed, the safety program manager and safety committee start setting goals and objectives for the program. This goal setting is accomplished through the common approach taught in many management courses and may be the organization's strategic planning for safety.

goals
broad statements of what needs to be accomplished

The first step in this process is to determine **goals.** Goals are defined as broad-based statements with a measurable outcome and time frame. Goals can be developed by the safety program manager alone or with the safety committee. However, using the group process and the safety committee will probably result in better acceptance by the members, and, generally, the group process provides a better result. There may be any number of goals set forth and these become the road map to guide the safety program. It should be cautioned that the number of goals developed for a particular time frame should be limited, and the goal should be realistic and obtainable. If there is too much to do in the specified time frame or if goals are never obtained, the organization's members may quickly lose interest and the program may suffer.

Note
The number of goals developed for a particular time frame should be limited, and the goal should be realistic and obtainable.

Using the case study presented at the beginning of the chapter, the safety committee had identified a particular problem with back injuries. After the problem was identified, the committee began goal development. After discussion, consensus was reached and the following goal recommended.

> Safety Program Goal 1: To develop a comprehensive back injury reduction program to reduce the number of job-related back injuries by 50% within the first 12 months of implementation with additional 25% reduction each 12 months thereafter.

■ Note
Whereas a goal is broad based, an objective is specific, but also must be measurable and within a given time frame.

In analyzing this goal, it is determined that the goal is broad enough, does not give any specifics, is measurable (50% reduction), and has a specified time frame (12 months). This would be a good goal statement for any back injury reduction program.

Once the goal statement has been developed, more specific statements of action must be developed in order to meet the goal. These specific statements, or objectives, provide the road map to reach a goal. Whereas a goal is broad based, an objective is specific, but also must be measurable and within a given time frame.

■ Note
An objective must only deal with one specific activity and, as with goals, be obtainable within the given time frame.

There is no magic number of objectives needed to reach a goal. Instead, it depends on the complexity of the goal and the number of interrelated activities needed. An objective must only deal with one specific activity and, as with goals, be obtainable within the given time frame. The following is a poorly written objective based on our fictitious back injury problem:

> Objective 1.1: A training program will be developed and presented to all employees, and each employee will be issued a back support belt within 30 days.

An analysis of this objective reveals a couple of flaws. First, the objective deals with more than one interrelated issue, training and support belts, and, second, the time frame set forth may not be reasonable. The four following objectives are written to comply with the suggested guideline and are designed to meet the back injury reduction program goal:

> Objective 1.1: Within 90 days, develop a standard operating procedure requiring assistance when performing nonemergency lifting of equipment over 50 pounds and train employees in the application of the procedure.
> Objective 1.2: Within 120 days, form a team to research options for relocating heavy equipment on the apparatus to lower levels and submit a written report with recommendations to fire department management.
> Objective 1.3: Within 120 days, consult with the department physician to develop a fitness routine designed to strengthen back muscles and incorporate education on proper lifting into daily fitness training.
> Objective 1.4: Within 180 days, research the use of back support belts in emergency service applications, including organizational results, and file a written report to the safety committee including recommendations and a cost-benefit analysis.

These objectives are all clear and to the point. They are measurable and include reasonable time frames. Each could be easily analyzed at the end of the specified time frames for completion.

The goal and objective process is dynamic and changing from two perspectives. First the goals and objectives have to be developed for each problem area and may change over time as new problems are identified. Second, as part of the process, each goal and objective should be reevaluated during the implementation process to see that the objective is being met, and if not, what has to be changed.

Note Safety program goals and objectives should be published and recognized by all members of the organization.

Safety program goals and objectives should be published and recognized by all members of the organization. This can be accomplished through the safety committee meeting minutes or any other departmentwide distribution process.

ACTION PLANNING

As described in the previous section, the goal statements are the road map for the safety program and the objectives are the road map to the goal. Then what is the road map to the objectives? The answer is the action plan (see Figure 10-1). The action plan is a step-by-step written guide to meeting an objective. Each objective should have an action plan developed.

The action plan has several components to be considered. The action plan should list the goal and objective that it is designed for and team member's names. The action plan should be developed by the team that has been assigned the objective and should contain very specific step-by-step actions, often set up in tabular form. For each step, the action, a completion time benchmark, the person responsible, the resources needed, any support or roadblocks anticipated, and a completion date should be examined and included. The only column that is sometimes left blank is the support/roadblock column, although generally support and roadblocks should be identified during the action planning process. Table 10-1 is an example action plan for objective 1.2 of our back injury reduction plan.

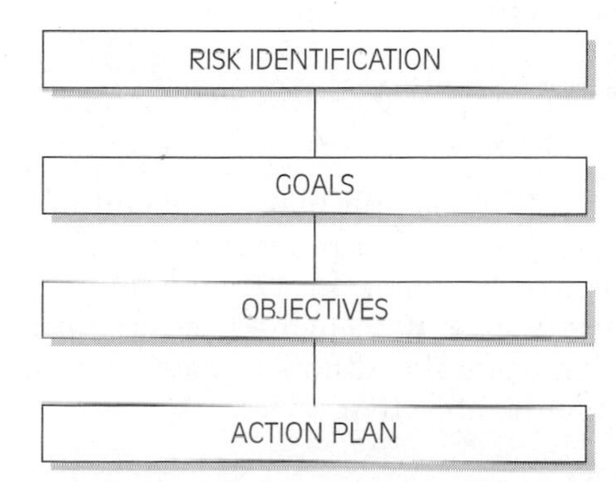

Figure 10-1 *The relationship between goals, objectives, and the action plan.*

Table 10-1 *Sample action plan.*

Safety Committee
ACTION PLAN

Date: January 1, 1999 **Team Members:** Jones, Smith, Big, Little

Goal 1: To develop a comprehensive back injury reduction program to reduce the number of job-related back injuries by 50% within the first 12 months of implementation with additional 25% reduction each 12 months thereafter.

Objective 1.2: Within 120 days, form a team to research options for relocating heavy equipment on the apparatus to lower levels and submit a written report with recommendations to fire department management.

Step No.	Action	Date To Be Completed by	Person Responsible	Resources Needed	Support(s)/ Roadblocks(R)	Date Completed
1	Review SOP adoption guidelines	1/10/99	Jones	None	None	
2	Contact other similar departments for SOPs	1/15/99	Smith	Telephone/E-mail Fax	(S) Chief association/ (S) Statewide labor organization	
3	Develop SOP	1/30/99	Smith	Computer	(S) Safety committee	
4	Determine heavy equipment	1/10/99	Big/Little	None	None	
5	Determine compartmentation options	1/30/99	Big/Little	Vehicle compartment inventory	(S) Company officers	
6	Have SOP draft approved	2/15/99	Smith	None	(S) FD Staff (R) FD Staff	
7	Begin training on SOP	2/20/99	Smith/Jones	Overhead training time	(R) Personnel acceptance	
8	Move equipment as recommended	3/1/99	Big/Little	None	(R) Personnel acceptance	
9	Full implementation	3/30/99	Team	None	None	
10	Evaluate outcome Recommend improvements	3/30/00	Team	None	(S) FD Staff	

As seen in Table 10-1, the action plan covers all of the necessary steps to fully implement the objective. Further progress can easily be measured throughout by using the completed benchmarks. Certain support and roadblocks have been identified so the team can anticipate these issues during the implementation process. This action plan example, although maybe not all inclusive, is a good example of a road map to the objective.

COST-BENEFIT ANALYSIS

Although the action plan presented in Table 10-1 did not include a cost-benefit analysis, very often one is necessary to support a position, particularly if the objective involves any financial outlay. A cost-benefit analysis is often used to show that the initial outlay for a program will save in future reduction of risk. Examples might include the purchase of bunker pants and the reduction in burn injury costs or the cost of an extensive driver training program and the associated reduction in vehicle accidents.

Often when using a cost-benefit analysis, some assumptions must be made and sometimes the data are based on estimations of improvement rather than actual results. Unfortunately it is almost impossible to determine the exact outcome of a program prior to its implementation. However, using information from similar agencies that have done similar projects, an estimation of expected outcomes can be formed. Another way to view outcome expectation is to analyze historical injury data and determine which would have likely been prevented if the program in question had been in effect. Another issue with cost-benefit analysis is that you cannot put a price on some losses that the emergency services incur. For example, the loss of one life is too much, and therefore a program would have to be undertaken regardless of the resultant cost-benefit analysis result.

A cost-benefit analysis typically allows the safety program manager to evaluate the cost-effectiveness of a program. The analysis examines the current cost of the risk and compares those costs to the cost of program implementation, considering both direct and indirect losses and costs. Recall from Chapter 3 that the cost of a risk is measurable and can be both direct and indirect. Direct cost might include the costs of medical treatment, the overtime paid to cover a vacancy on a crew, or the cost of replacing equipment. Indirect costs include loss of productivity, the loss of using the equipment, stress-related concerns of coworkers, and possibly the cost of replacing the employee. Direct and indirect costs may also be applied as program costs.

■ Note

The first step in the cost-benefit analysis is to describe, numerically, the cost of the risk currently.

The first step in the cost-benefit analysis is to describe, numerically, the cost of the risk currently. Again, using our back injury example, a study can be performed and can describe what is occurring at the present time, without intervention. The next step is to determine the cost of the risk after the intervention has been implemented. After this step, the manager can determine whether this measure would be effective from a purely cost standpoint. Finally, the cost of the program implementation has to be calculated. Once these three areas have been calculated, an informed decision can be made. Remember that a program that may only save $1,000 per year and cost $5,000 to implement, may still be a good program because of the organizational benefit after the fifth year. Figure 10-2 is an example spreadsheet using the back injury reduction example. The analysis is for illustration only and is not to be construed as showing exact costs.

An examination of Figure 10-2 reveals that the organization would save $12,180 the first year. Aside from some ongoing training and replacement back

Cost/Benefit Analysis	
Goal: Development of Back Injury Prevention Program—First Year	
Current Situation and Costs	
25 Back injuries per year with resultant hospital/Doctor visit average $1000 per visit	
and an average 1.5 (36 hours) days of work lost	
Direct Costs	
Medical Expenses	$25,000
Indirect Costs	
Overtime to Cover Vacancies on Shift Average Hourly Rate $10.00	
36 × 25 × 1.5 × $10.00	$13,500
Total Costs	$38,500
Future Estimation of Situation	
13 Back injuries per year with resultant hospital/Doctor visit average $1000 per visit	
and an average 1.5 (36 hours) days of work lost	
Direct Costs	
Medical Expenses	$13,000
Indirect Costs	
Overtime to Cover Vacancies on Shift Average Hourly Rate $10.00	
36 × 25 × 1.5 × $10.00	$ 7,020
Total Costs	$20,020
First Year Savings (Estimation based on research)	$18,480
Program Costs	
Direct	
Physicians Time	$ 2,000
Back Supports	$ 2,500
Indirect	
Training Time	$ 1,000
Apparatus Compartmentation Change	$ 800
Total Program Costs	$ 6,300
Total Cost/Benefit—First Year	
First Year Savings minus Program Implementation Costs	$12,180

Figure 10-2 *An example cost-benefit analysis of a back injury prevention program.*

supports, the cost of the program in future years would be greatly decreased. Therefore, the savings over time would be higher and could be calculated using the same format.

TRAINING

There is a direct and close relationship between training and safety (see Figure 10-3). In fact, in many organizations the training officer or division has assumed the safety program functions. Many of the training programs offered have a basis in operating safely. Many of the safety initiatives that a department undertakes are disseminated through the department's training programs. From the back injury program example, a key component involved the training of members.

■ **Note**

Training mandates are designed in hopes of ensuring that a minimum level of training is provided to responders in order that the responders can handle the incident safely and recognize dangers.

As presented in Chapter 2, a number of standards and regulations dictate how an emergency service organization operates. Within these standards and regulations are requirements for certain levels of training. A very good example is the hazardous materials waste operations regulation, which bases training requirements on five levels of response. Training mandates are designed in hopes of ensuring that a minimum level of training is provided to responders in order that the responders can handle the incident safely and recognize dangers. Many states require a minimum level of training for firefighters or paramedics. Within these training requirements, one can always find some requirement related to safety, whether it be use of a self-contained breathing apparatus (SCBA) or infection control.

Figure 10-3 *A training class on fire extinguishers.*

Training is also important as the organization's safety program develops. The training program can be a vehicle for introducing and testing new procedures. Remember, the best accountability system on paper that does not work during training evolutions will probably not work at a real incident. Training may also be developed to help deal with an existing injury problem, such as lifting injuries or thermal burns.

More than any other function within an emergency service organization, the training staff should be the most integrated into the safety program, both in development and in implementation.

DEVELOPING STANDARD OPERATING GUIDELINES AND SAFETY POLICIES

The development of safety procedure and polices is necessary in order to meet some of the goals and objectives defined in the program. After development and approval, they must be reviewed for effectiveness and updated as necessary.

In starting the process, the committee developing the SOP or policy should review the applicable goal or objective that the policy has to satisfy. Other organizations may be contacted for copies of their SOP for the same objective. A gap analysis can also be performed, which simply answers the question of where we are versus where we should be.

Once these few steps have been completed, the committee can develop the SOP based on the organization's format. It is important that, once a draft is developed, the draft is sent out for comment. It is a good idea to get comments from other members of the department and from various ranks within the organization. If the SOP or policy is specific to one work function or one workplace it is very important to target the affected groups. This inclusion allows for greater input and help with member buy-in once adopted. Getting feedback may also tell the committee that the procedure may not work in real application.

Once this entire process has produced an approved new or changed SOP or policy, all members of the organization should be trained in its application. If practical, there should be realistic training, not just a classroom reading. Further feedback should be provided to ensure the procedure will work.

Summary

In order to be effective, the safety and health program must have several essential elements including a top management that is committed to the program, a safety and health program manager, some type of record-keeping system for data analysis, an incident safety officer or officers, a training program, standard operating procedures, proper equipment and PPE that meets requirements and standards, a safety committee, a department physician, and access to local, state, and national injury and death statistics. Some of these items may come after program development. However, at a minimum the program should have the commitment of top management, a safety program manager, a safety committee, a record-keeping system, and access to data.

Once risks are identified and analyzed, the safety program manager, in conjunction with the safety committee, must develop goals and objectives designed to minimize the risks. Goals are broad-based statements that are measurable and realistic and provide a guide to the safety program. Objectives are more specific and are the guide to meeting the goal. There may be one or many objectives for each goal. An action plan is developed based on each objective. The action plan is a step-by-step guide to meeting the objective. For each step in the action plan—the action, a completion time benchmark, the person responsible, the resources needed, any support or roadblocks anticipated, and a completed-by date—should be examined and included.

For most goals and objectives, a cost-benefit analysis should be performed. The cost-benefit analysis examines the current situation in terms of direct and indirect cost. The costs of the risk after program implementation and the expected cost of the program are compared to the current cost, so that an informed decision can be made regarding implementation.

Training and safety are very closely related and somewhat dependent on each other. In some organizations the training division actually does the safety function as well. Many training programs are designed for educating the members on how to perform tasks safely. Development of safety- and health-related SOPs and policies are also integral to the program. These can be developed by the safety committee after gap analysis, review of the safety goal and objectives, and after contact with other organizations with similar policies. Once a draft is developed, it is important to get feedback from all levels of the department and to train all members on the implementation.

Concluding Thought: The safety and health program has several components. Using these components, the safety program manager goes through various processes to implement the program.

Review Questions

1. A department physician is an essential component to have in place before a safety program can be established?
   A. True
   B. False
2. A goal is a broad statement with measurable results?
   A. True
   B. False
3. An action plan can be considered the road map to a goal.
   A. True
   B. False
4. List two ways the training program can be of benefit and works in conjunction with the safety program.
5. Which of the following is applicable to a cost-benefit analysis?
   A. The cost of the safety gear to be purchased.
   B. The current direct cost of the injuries if the program were not introduced.
   C. The indirect costs associated with the program after implementation.
   D. All of the above information would be necessary.

Activities

1. Select an injury problem within your department. Write one goal and at least two objectives to develop a program to reduce or eliminate the problem.
2. Write an action plan for each objective.
3. Perform a cost-benefit analysis on the program.
4. Review your department's safety program and the relationship to the training division. Are they working together? Do they have similar goals? Is it effective?
5. Review your department's SOP development process. How does it compare to that presented in the text?

Safety Program Evaluation

Learning Objectives

Upon completion of this chapter, you should be able to:

- Describe the purpose for evaluation of the health and safety program.
- Compare the two types of evaluations, process and outcome.
- Explain who has the responsibility for evaluation.
- Describe a recommended frequency for evaluation and the factors that affect the frequency.

CASE REVIEW

The large urban emergency medical services (EMS) agency had placed a comprehensive injury reduction program into effect more than 3 years ago. The program was evaluated annually using outcome evaluation. This evaluation process examined such areas as injury rate and severity, attitude changes among personnel, a knowledge change in the personnel regarding injury prevention, and an evaluation of how well the policies were being complied with.

The first annual evaluation revealed that the injury reduction program was, in fact, meeting its goal of reducing injuries. In the first 12 months, the injury rate dropped by 50%. Subsequent annual evaluations showed similar changes in the injury rate, however injury severity showed to be on the increase. The outcome evaluation focusing on the foregoing areas revealed that during the first years of the program, the policy changes were complied with very well throughout the department. However, due to a number of new employees hired over the subsequent 2 years and a number of promotions, many of the policies associated with the program lacked priority in terms of compliance.

Although the number of injuries continued to decrease, the severity increased as employees were failing to follow established procedures. It is imperative in program evaluation that all components be evaluated and evaluated from several different dimensions.

INTRODUCTION

■ Note **The method of evaluation should compare where we are now to where we were prior to program implementation.**

No health and safety program is complete without some sort of evaluation process. Once the program has been designed and implemented, a process must be undertaken to evaluate its effectiveness. This evaluation may take any number of forms, depending on the goal or objectives being evaluated. Furthermore, the time frame for evaluation will also be goal dependent. Simply put, the method of evaluation should compare where we are now to where we were prior to program implementation. In order to be effective, an understanding of the evaluation process and the measuring of results must be undertaken. Once this knowledge is gained, then the process can be included in the program (see Figure 11-1).

THE EVALUATION PROCESS

The process of evaluation may take different forms, however it usually involves the comparison of statistics from one time period with another time period. The results can then be compared to what was expected. Differences between expected outcomes and actual outcomes should be evaluated for program changes. There are three reasons to evaluate the program:

1. To see if the program is effective
2. To determine the response to the program from the members' perspective
3. To facilitate program changes

Figure 11-1 *A safety program manager analyzes data for the annual evaluation.*

process evaluation
evaluation of the various processes associated with a program or task that is ongoing

outcome evaluation
an evaluation that answers the question, "Did the program meet the expected goals?"

There are two methods of evaluation that should be used: **process evaluation** and **outcome evaluation.** Process evaluation is an analysis of the procedures of the program and is undertaken throughout the program. Basically, the process evaluation answers the question, "How well did the program do what it was intended to do?" For example the back injury prevention program designed in Chapter 10 had a component involving the training on proper lifting techniques. A process evaluation for this component would examine the back injuries caused by lifting in a given period and determine if the training had an impact.

Outcome evaluation is measured after the program has been in effect for a while. This evaluation examines what the program did. Again, using the back injury prevention program as an example, an outcome evaluation would be used at the end of a time period, probably a year, to determine what effect the program had. Several areas should be analyzed, including:

- The injury rate
- The severity of injuries
- An attitude change among members regarding the back injury problem
- A knowledge change among members regarding proper lifting and the back injury problem
- Compliance with policy changes

There are several components involved in both the process and the outcome evaluations.

Process Evaluation

- Determine who was affected by the program.
- Determine to what extent they were affected.
- Are improvements occurring as planned?
- What parts of the program appear to be most effective?
- Which parts of the program appear to be least effective?
- What program changes are needed to increase the effectiveness of the program?

Outcome Evaluation

- Examine the current injury rates and severities, comparing them to those prior to program implementation (using the goals and objectives as benchmarks).
- Measure the change in knowledge and attitude.
- Measure behavior changes.
- Analyze the changes in the physical environment.
- Measure the response to policy changes.

cognitive
skills learned through a mental learning process as opposed to practical learning

The safety program evaluation should include both of these evaluation methodologies. Much of the information needed can come directly from the injury statistics, however, there will be other required measurements. For example, when analyzing the change in attitude and knowledge, a more **cognitive** measurement may have to be used.

RESPONSIBILITY FOR EVALUATION

Who has the responsibility for program evaluation? The answer should be "Everyone." Everyone can and should play some role in the evaluation process. The safety program manager may oversee an entire outcome analysis while the responders on the street perform a process evaluation each time they respond to an alarm.

Ultimately, the safety and health program manager has the responsibility to see that evaluation gets done. However, this may be accomplished through the safety committee, incident safety officers, senior staff members, first-line supervisors, and individual members.

The safety committee can assist with the compiling and analysis of injury data and also help with the determination, through feedback, of knowledge and attitude changes. The incident safety officers are valuable in providing feedback during process evaluation and for measuring compliance with new policies or procedures. Senior management will be part of the process by making resources and data available to the safety and health program manager and the safety committee. First-line supervisors also will be feedback providers and can very often measure knowledge and attitude changes. First-line supervisors also are integral for feedback on policies and procedure effectiveness. Finally, the individual member should be empowered to make suggestions to improve program components and can measure peer attitude and knowledge changes.

Although not cost-effective to do each year, the NFPA 1500 standard requires external evaluations, which might involve hiring a consultant in the safety and health profession or using health and safety program managers from other departments. Workers' compensation and other insurers have risk managers on staff and often make these experts available to a client for an external look at the safety program.

Having an external evaluation performed is a positive effort for both the safety program manager and the program in general. Very often, a person from the outside who is not close to the issues will find areas in which improvements can be made that may not be as obvious to the person on the inside.

EVALUATION FREQUENCY

When the program should be evaluated is another question. As mentioned previously, the two types of program evaluation, process and outcome, can help to answer this question. Process evaluation can almost be a continuous process. Outcome evaluation should occur at the end of a specific period of time, usually one year or whenever the program goal or objective defines as a measured time frame.

The frequency of evaluation should be considered dynamic. For example, if a serious injury or death occurs, an immediate evaluation or analysis into the safety program components dealing with the cause should be undertaken right away. If new technology is introduced into the profession, an analysis into the applicability to the program should be examined.

Program evaluation is a required component of the safety and health program. In fact, program evaluation actually occurs before the program starts; risk identification and analysis are types of evaluations. They just happen to be evaluations of a program before a program begins.

Summary

There are two basic formats for the evaluation of the program, the process evaluation and the outcome evaluation. The process evaluation should be viewed as an ongoing analysis into the program to determine whether the program is reaching the intended recipients and to what degree. It also should reveal what changes have occurred in knowledge and attitude. The outcome evaluation should occur after the program has been in effect for some period of time and actually measures results. These results can be compared to the program goals and objectives and the program effectiveness can be determined.

Everyone in the organization has some role in the evaluation process. Although the responsibility rests with the safety and health program manager, the safety committee, incident safety officers, senior staff, first-line supervisors, and the members all have a role in the evaluation process.

Although a process evaluation is ongoing and the outcome evaluation is after a defined period of time, the frequency of evaluation is dynamic. If a major event occurs in the organization or if new technologies are introduced, a program evaluation is warranted at that time. After the evaluation, regardless of the type or the frequency, the information gained should be used for program enhancement and improvement.

Concluding Thought: A safety and health program cannot be successful without evaluation methods as part of the process.

Review Questions

1. List the two types of program evaluations.
2. List the components required for each of the two program evaluations listed in Question 1.
3. List three purposes for program evaluation.
4. What is the role of the first-line supervisor in program evaluation?
5. Describe considerations for determining the frequency for evaluation.

Activities

1. Compare the evaluation of the safety program in your organization to that presented in the text. How does your process compare?
2. What changes do you recommend?
3. Is the role of the various members of your organization similar to the recommendation of the text in terms of program evaluation? If so, is it effective? If not, is it effective?

Information Management

Learning Objectives

Upon completion of this chapter, you should be able to:

- Describe the purpose of data collection and reporting.
- Identify the data that should be collected within the organization.
- Identify the data that should be collected for outside organizations.
- Describe the purpose and process for publishing an internal health and safety report.
- Describe the use of the Internet as a health and safety information source.

CASE REVIEW

The safety and health program manager was new in the job, having been assigned only 2 months before. The emergency services director sent a memo outlining concerns about bloodborne pathogen exposures. The director felt that the current procedures and policies within the department were not adequate for the protection of the employees.

The safety program manager sought to examine the current system, review the national experience, and find up-to-date information from the electronic media. First, the exposure records for the organization were analyzed. This was accomplished by doing a computer search of the injury database, searching for records that listed exposure for type of injury. The manager then contacted the International Association of Fire Fighters (IAFF) and requested a copy of the most recent injury and exposure report. The information from these two reports were compared and the determination made that the organization's level of exposure was greater than the national average.

Having information in hand, the safety program manager filed a report with the director that explained where the organization was and where it should be. The director gave the OK to develop and implement changes to the existing program.

Wanting to know more about the problem and to look at how other agencies were handling the situation, the safety program manager accessed the Internet. Performing a search on the words *bloodborne*, *pathogen*, and *exposure*, the Internet search engine provided the manager with more than 100 sites to visit that would have information on exposures. Some of these were the Department of Labor, which provided a copy of the Occupational Safety and Health Administration (OSHA) standard on bloodborne pathogens; the Centers for Disease Control, which had a copy of the Ryan White law and statistics dealing with the disease; and the United States Fire Administration (USFA), which had documents and sample policies from numerous fire departments across the country.

With the internal and external data available and the information from the Internet, the safety program manager had the resources necessary to develop improvements and implement a new plan.

INTRODUCTION

■ Note
In order to be useful, data must be retrievable in a format that is compatible with its intended use.

Like program evaluation, data collection and information analysis is a critical component of the health and safety program. Data collection and analysis is used for many of the other program components discussed, including the development of goals and objectives and the evaluation of the program or components. In order to be useful, data must be retrievable in a format that is compatible with its intended use. Although not always possible, this process should be facilitated through the use of a computer.

Data is collected for use within the organization. Data is also collected for and by outside agencies and used for various other purposes. The collection and analyzation by outside agencies is helpful for local comparison to national trends. Once the internal data is collected and analyzed, an annual health and safety report should be published for all levels of the organization.

The environment is rapidly changing and so is the speed in which we can communicate. Our access to information has never been greater. The safety and health program manager can use the Internet to obtain volumes of information regarding safety and health, regulations and standards, and copies of policies from other similar departments regarding the safety and health program.

PURPOSE OF DATA COLLECTION AND REPORTING

There are many reasons for performing data collection. For starters, data collection and reporting is required for many jurisdictions by OSHA or a state department of labor. It is equally important for the organization to have the information available for analysis and evaluation. Furthermore, good record keeping helps to protect the organization and the members from a legal standpoint (see Figure 12-1).

Records regarding workers' compensation and other insurance claims can help predict increases or decreases in premiums that will have future budget impacts. Further, the mandatory reporting requirements of federal and state agencies usually carry fines associated with noncompliance.

Bear in mind that most, if not all, medical-related data and information is considered confidential and the access to the information should be kept to only those with a need to know. Department standard operating procedures (SOPs) should include who has access to medical-related data.

INTERNAL DATA COLLECTION

Internal data collection is that associated with health and safety data that is generated from within the organization and is used within the organization. Several common data types fall into this category, including:

- Injury reports
- Accident reports
- Individual medical histories
- Drug-free workplace test results
- Reports dealing with an employee killed in the line of duty
- Exposure records

In order to collect the data required for these reports, standardized forms are often used to collect the data. The data should then be placed into a computer database for storage and later retrieval. Safety and health program managers should check their local, state, and federal laws for requirements in keeping hard copies of these records and for the length of time they must be maintained. Regardless of the requirements, the information should still be filed electronically, in addition to required hard copies for easy retrieval and analysis. Figure 12-2 is an example of an accident/injury reporting form with the generally required information.

SOUTHWEST FLORIDA PROFESSIONAL FIREFIGHTERS
LOCAL 1826 / I.A.F.F.
EXPOSURE REPORT

1. Firefighters Name ____ 2. Social Security # ____ 3. 1826 Dist. # ____
4. Incident Date ____ 5. Alarm Time ____ 6. Incident # ____

A MEDICAL / RESCUE EXPOSURE

7. Patients Name ____
8. Facility Patient Transported To ____
9. Transporting Agency ____
10. Name of Suspected or Known Disease: ____
11. Precautions Taken By You at Scene and After Exposure
☐ gloves ☐ face mask ☐ goggles ☐ protective gown
☐ washed exposed skin ☐ physician treatment (if this is checked complete *section F*)

How you were exposed	A. Topical Skin Contact	B. Needle Puncture	C. Direct contact to open wound	D. In Eyes	E. In nose or mouth
12. BLOOD					
13. URINE					
14. SPUTUM					
15. VOMITUS					
16. FECES					
17. OTHER					

B FIRE INCIDENT TYPE

18. ☐ Residental Fire 19. ☐ Industrial Fire 20. ☐ Vehicle Fire 21. ☐ Commercial Fire 22. ☐ Wildland Fire 23. ☐ Trash/Dumpster fire 24. ☐ Marine Fire 25. ☐ Explosion 26. ☐ Rescue 27. ☐ Spill
28. Other (describe in 1 or 2 words) ____ 29. Type of occupancy (1 or 2 words) ____

C LENGTH OF EXPOSURE BY FIRE STAGE / ACTIVITY

FIRE STAGE :	A <1 hr.	B 1-2 hr	C 2-3 hr	D 3 + hrs
30. Incipient				
31. Free Burning				
32. Smoldering				
33. Non-Fire Incident				

	A < 1 hr	B 1-2 hr	C 2-3 hr	D 3 + hrs
34. Extinguishment				
35. Empty / Ventilation				
36. Rescue				
36. Light Overhaul				
37. Heavy overhaul				
38. E.M.S.				
39. Investigation				

D SMOKE / CHEMICAL/ OTHER EXPOSURE

41. Smoke Conditions ☐ light ☐ Heavy ☐ None
42. Smoke Color ____

Chemicals Present	A Vapor Gas	B Dust	C Liquid Heavy Mist	D Light Mist	E Smoke	F Solid Powder
43.						
44.						
45.						
46.						

47. Type of Chemical Exposure to Above: ☐ Inhaled ☐ Ingested ☐ Skin Contact ☐ Eye Contact ☐ Other ____

E SYMPTONS

	AT INCIDENT	SYMPTON	AFTER INCIDENT	FOR HOW LONG
48.		Eyes Burn		
49.		Cough		
50.		Nose Bleed / Cough Blood		
51.		Nose/Lung Irritation		
52.		Nausea/Queasiness		
53.		Dizzy		
54.		Ears Ringing		
55.		Headache		
56.		Skin Rash/Burn		
57.		Unconscious		
58.		Other:		

F MEDICAL DIAGNOSIS

59. DID YOU RECEIVE MEDICAL EVALUATION OR TREATMENT AFTER EXPOSURE? ☐ YES ☐ NO
60. DIAGNOSIS ____
61. DOCTORS NAME ____ 62. FACILITY ____

G SPECIAL EQUIPMENT / DECONTAMINATION ☐ PASS ALARMS ☐ BUNKER PANTS ☐ HELMETS ☐ BUNKER COATS ☐ GLOVES ☐ HOODS ☐ SCBA'S ☐ BOOTS

63. WERE YOU SUPPLIED SPECIAL EQUIPMENT FOR THIS INCIDENT? ☐ YES ☐ NO DESCRIBE ____
64. WERE SPECIAL DECONTAMINATION PROCEDURES FOLLOWED AFTER THE EXPOSURE ☐ YES ☐ NO DESCRIBE ____

H CO-WORKERS AT TIME OF EXPOSURE

65. PLEASE LIST NAMES OF OTHER FIREFIGHTERS WORKING CLOSE TO YOU AT TIME OF EXPOSURE ____

I ADDITIONAL INFORMATION (YOUR FUNCTION AT INCIDENT, UNUSUAL CIRCUMSTANCES, ETC.)

66. ____

J 67. DATE REPORT FILLED OUT ____ 68. NAME OF PERSON FILLING OUT REPORT (PRINT) ____

69. SIGNATURE ____

Used by permission of Local 1826 IAFF.

Figure 12-1 *An example of an exposure-reporting form.*

Report #____________

Accident / Injury / Investigation Report
PRINT ONLY

TO BE FILLED OUT BY INJURED PARTY ***Today's Date:***____________

Name of Injured:	Sex: Male ❑ Female ❑	Age:
Date & Time of Accident: AM PM	Location of Accident:	

Describe the accident and how it occurred:
Was accident reported to immediate supervisor? Yes ❑ No ❑ / Date & Time reported to Supervisor:
Were there any injuries? Yes ❑ No ❑ If yes, describe injuries:
Was medical treatment given? Yes ❑ No ❑ If yes, give physician name & address:
Was time lost from work? Yes ❑ No ❑ If yes, explain.
Were department SOP's adhered to? Yes ❑ No ❑ If no, explain.
Were safety standards adhered to? Yes ❑ No ❑ If no, explain:
Was law enforcement notified? Yes ❑ No ❑ If yes, which law enforcement agency?
What could have been done to avoid this accident?
SIGNATURE OF INJURED PERSON:

TO BE FILLED OUT BY IMMEDIATE SUPERVISOR

Cause of accident:		
Witnesses to accident? Yes ❑ No ❑ If yes, list witnesses		
Was personal protective equipment required? Yes ❑ No ❑ Was it being used? Yes ❑ No ❑ If NO, explain:		
Interim corrective actions taken to prevent recurrence:		
Permanent corrective action recommended to prevent recurrence:		
SUPERVISOR SIGNATURE	**DATE:**	
Status and follow-up action taken by Safety Officer		
SAFETY OFFICER SIGNATURE	**DATE:**	**REVIEWED BY CHIEF / DATE**

Figure 12-2 *Sample accident/injury-reporting form. Courtesy Palm Harbor Fire Rescue.*

EXTERNAL DATA COLLECTION

External data is the information, collected internally, that is to be used by another agency outside of the organization. This is often state and national databases, workers' compensation carriers, and insurance companies. As might be recalled from Chapter 1, the IAFF/IAFC joint labor management wellness/fitness initiative requires reporting to a national database that is accessible by local organizations.

The information that is required for the external agencies may or may not be the same as that needed within the organization. However, usually the internal use forms provide adequate background information for the external reporting requirements.

Workers' Compensation

Clearly, the organization's workers' compensation carrier is one external organization that wants injury data. Generally, there are rules and laws that provide for specific time frames in which a workers' compensation carrier must be notified. The workers' compensation carrier will also, upon request, provide the organization with a summary of claims over a requested period of time. This can be a useful tool in program evaluation. For example, knowing that internally the number of workers' compensation claims have gone down is good news. However, if the cost of the claims, in terms of lost time and treatment, have increased, the program will have to be reevaluated and changes implemented.

Occupational Safety and Health Administration

Although OSHA does not apply to every public fire department in the country, there are reporting requirements in the states where OSHA does apply. OSHA requires that the **OSHA 200** log be posted annually in the workplace for all to see. This is to ensure that the organization is, in fact, keeping records on injuries throughout the year. However, there are guidelines on what types of injuries must be reported. An injury must be reported if medical treatment more than first aid is received, if the injury results in job loss or transfer, or if the injury resulted in a period of unconsciousness.

OSHA 200
a list of all on-the-job injuries within a given work site; required to be posted for all to see

States that do not fall under OSHA for public emergency response agencies should contact their state Department of Labor to determine a state requirement for similar reporting.

National Fire Protection Association

As described in Chapter 1, the NFPA publishes an annual injury report and an annual firefighter fatality report. In order for these reports to reflect a national experience, organizations are asked to participate in their data collection program.

Remember that the injury portion of the NFPA's reports are a sampling designed to predict the national experience.

United States Fire Administration

The USFA also does health and safety data collection thorough the Nation Fire Incident Reporting System (NFIRS). Although voluntary, the NFIRS has components for reporting firefighter casualties, both fatalities and injuries. This information is reduced to an annual report that addresses the nationwide experience for those organizations that participate.

International Association of Fire Fighters

Like the NFPA and OSHA, the IAFF collects data for annual injury, exposure, and fatalities reports. The data collected by the IAFF is only from paid fire departments with IAFF affiliation and includes the United States and Canada. The published report considers EMS-related activities to a greater degree than the previous two reports.

National Institute for Occupational Safety and Health

The National Institute for Occupational Safety and Health (NIOSH) began a project in 1997 in which it will investigate firefighter line of duty deaths. The project, called Fire Fighter Fatality Investigation and Prevention Program will be funded in 1998. Text Box 12-1 describes this program.

TEXT BOX 12-1 DESCRIPTION OF FIRE FIGHTER FATALITY INVESTIGATION AND PREVENTION PROGRAM (COURTESY NIOSH).

In fiscal year 1998, Congress recognized the need for further efforts to address the continuing national problem of occupational firefighter fatalities, and funded NIOSH to undertake this effort. The Congressional language states in part:

> In FY 1998, $2.5 million will be needed to conduct fatality assessment and control evaluation investigations to gather information on factors that may have contributed to traumatic occupational fatalities, identify causal factors common to firefighters' fatalities, provide recommendations for prevention of similar incidents, formulate strategies for effective intervention, and evaluate the effectiveness of those interventions.

In brief, the overall goal of this program is to better define the magnitude and characteristics of work-related deaths and severe injuries among firefighters, to develop recommendations for the prevention of these injuries and deaths, and to implement and disseminate prevention efforts. A five-part integrated plan, centered around the field investigation of firefighter fatalities, is outlined below. This plan will remain flexible, as new staff are hired and become familiar with the critical hazards and many related issues in firefighting and as results from field investigations become available.

I. NATIONAL FIRE FIGHTER FATALITY INVESTIGATION PROJECT

This is the cornerstone of the overall NIOSH program to prevent firefighter line-of-duty fatalities. The objectives for this effort include the investigation of all occupational firefighter fatalities to assess and characterize the circumstances of these events in order to develop succinct descriptive and evaluative reports for distribution to the fire community across the country. This work will be carried out by the NIOSH Fire Fighter Investigation Team. It is expected that the reports alone will have a major impact by better defining the causal factors of firefighter deaths, calling national attention to the problem, and providing insights into the prevention efforts that are needed. As noted below, several interrelated projects will enhance both the use and impact of this core effort.

This project has two major parts, the first involving NIOSH staff in the Division of Safety Research (DSR) and the Division of Respiratory Disease Studies (DRDS), and the second involving the Division of Surveillance, Hazard Evaluations, and Field Studies (DSHEFS).

A. Injury Fatality Investigations

DSR will use its Fatality Assessment and Control Evaluation (FACE) model for investigating occupational injuries to conduct investigations of all fireground and nonfireground fatal injuries. These investigations will include both career and volunteer firefighters. In addition, staff at the DRDS Respirator Certification and Quality Assurance Branch will assist with selected investigations in which the function of respiratory protective equipment may have been a factor in the incident. The DRDS laboratory and field staff will evaluate the performance of self-contained breathing apparatus as needed in this effort.

B. Cardiovascular Disease Fatality Investigations

Cardiovascular disease (CVD) has been a significant cause of work-related death among firefighters for many years. This DSHEFS activity will conduct multifactorial assessment of personal, physiological, psychological, and organizational factors associated with CVD deaths among firefighters while on duty. Directly related efforts will be considered, such as measuring carbon monoxide and other exposures and body burdens at various types of fires. The goal is to investigate most, if not all, such deaths, and to produce reports similar to those in Part A above.

II. FIRE FIGHTER FATALITY DATABASE PROJECT

It is expected that the individual firefighter fatality investigations and reports described previously will be very useful to other fire departments in their health and safety programs. However, the information will have additional use when it is organized and made easily accessible. Such a database will allow a multitude of analyses on topics including time trends, consistent but perhaps poorly recognized risk factors, various correlations, annual reports, and so forth. This project will use statistical and epidemiologic expertise, primarily based in DSR, to establish and maintain an electronic database using information obtained from the fatality field investigations noted previously. It will involve a number of issues, particularly in the initial setup, including the quality and completeness of the data, the development of appropriate and consistent investigation data collection instruments, data field decisions, data coding, data entry and checking, analysis protocols, and access procedures.

III. INTERVENTION RESEARCH PROJECT

This project will start slowly, as NIOSH gains experience and expertise in the causal factors and critical issues involved in firefighter fatalities. However, the purpose of the overall NIOSH program is to prevent such fatalities, and it is recognized that this effort has the benefit of many years of past data and research by a number of excellent private and public organizations and individuals. There may thus be selected issues that have moved beyond the surveillance and causal factor analysis stage to one in which an evaluation of an intervention is needed. NIOSH will pursue such issues with the goal that they may lead to the widespread implementation of life-saving practices now, rather than later after more lives are lost. This effort will be based largely but not exclusively in DSR, and will use scientific methods to determine whether specific equipment, practices, and programs are effective in reducing firefighter fatalities.

IV. APPLIED LABORATORY AND FIELD RESEARCH PROJECT

NIOSH will consider funding applied laboratory and field research projects on a case-by-case basis, developed in response to questions and issues raised by the field investigations, and by the fire community based on its experience. These projects will be specific, time-limited efforts rather than an ongoing funded program, and will be subject to available monies after higher priority projects are supported. All such work will be a collaborative effort in which NIOSH staff are heavily and continuously involved. Various organizations and individuals may be selected to participate in these projects in order to derive the greatest benefit.

In fiscal year 1998, NIOSH plans to address the substantial concern over firefighter protective clothing by entering into an interagency agreement with the National Institute for Standards and Technology (NIST) to conduct full ensemble personal protective clothing testing under actual fire conditions. NIOSH staff will be involved with all phases of the study, and fire community input will be solicited.

Future research efforts could involve NIST, academic researchers, and other appropriate organizations to address issues identified by the field investigations and by the fire community based on its experience and needs. This research is intended to complement NIOSH field investigation activities.

V. INFORMATION DISSEMINATION PROJECT

Appropriate information dissemination is essential if the NIOSH Fire Fighter Injury Prevention Program is to reach its goal. The major potential benefit of an investigation will not be to the involved fire department, which after a fatality often will understand the problems and needed changes, but to the other 36,000 fire departments and their over one million firefighters who need to address similar problems to prevent a similar tragedy. This national program to prevent firefighter fatalities and serious injuries will result in substantial amounts of new information that will be important to firefighters, researchers, program planners, and others. It is imperative that new and existing information be readily accessible to those who can use it for the development of practices, recommendations, and guidelines to prevent firefighter injuries and deaths.

Unfortunately, simply sending information does not always lead to the desired changes in behavior and practices; often complex social, economic, educational, and psychological issues are involved. Thus there are important questions regarding what types of products to create, what to send, how much, to whom, and by what avenues. This is another area that will need to be developed, and to be modified as the program progresses.

NIOSH plans to make the investigation reports, including a summary page and the full report, available via a World Wide Web home page that can be accessed either directly or through the NIOSH home page. It is considering the most effective way to distribute paper copies of the full reports and/or their summary pages. Additional issues relate to the best format for access to the electronic database noted previously, and distribution of periodic reports, such as an annual summary, describing time trends and other information. It is planned that these data will reach professional and lay journals as well as other media. Finally, the best approach to using regional and national conferences, training forums, and other organizations to disseminate both investigation results and recommendations will be explored. Because these types of activities are already being done by many individuals and groups, extensive networking will be used to effectively leverage the NIOSH efforts.

PUBLISHING THE HEALTH AND SAFETY REPORT

Once a data collection and retrieval system is put into place, the measurable data must be turned into usable information. Normally, this requires some sort of analysis and the compiling of a report over time, usually annually (see Figure 12-3). Once published, this report should be distributed to all levels of the organization, including elected officials and senior city or county officials. This report can help as an educational tool when support for resources is needed to

Figure 12-3 *The annual injury report should be distributed throughout the organization.*

enhance the program. The annual health and safety report should contain the following information, keeping in mind, again, the confidentiality of medical records:

- Introduction
- General state of the organization in terms of safety and health
- Accomplishments/improvements/benchmarks of the safety and health program for the reporting time period
- Goals and objectives for the next reporting period
- The analysis of the injury and fatalities for the reporting years
- A comparison to the national experience
- A comparison to similar departments in similar geographic areas
- A report on significant incidents, including findings and changes made for preventing future occurrences
- A summary of the organization's compliance with regulations and standards
- Other plans for improvement and resources requested
- A summary
- Specific graphics relating to the presentation of the information

Sometimes the health and safety officer will be called upon to officially present this report to the senior management or the elected officials. In such cases, the presenter should have adequate audiovisual aids to assist in this process, which might include slides, overheads, or computer generated graphics, charts, bulleted goal and objective lists, and sometimes pictures.

ACCESSING HEALTH AND SAFETY INFORMATION USING THE INTERNET

Note

The Internet, a network that interconnects computers worldwide, can provide access to other organizations' information, access to state and federal regulations, and access to annual national reports, such as those available through the United States Fire Administration.

We are living in the information age. With electronic mail, fax machines, and video conferencing, communication is more and more becoming instantaneous. Access to information continues to grow. The safety and health program manager, through the use of a personal computer, can have access to a great deal of information to assist with the safety and health program. Through the use of an organization's e-mail system, accident and injury reports can be filed more timely, and policy changes can be distributed to all work locations with the touch of a computer key. Granted not all organizations will have access to some of these systems, but for those that do, the amount and the quality of the information is invaluable.

The Internet, very simply, a network that interconnects computers worldwide, can provide access to other organizations' information, access to state and federal regulations, and access to annual national reports such as those available through the United States Fire Administration.

In order to access the Internet, an organization must have an Internet provider. The provider normally provides a local number for dial-up access and,

search engines programs on the Internet that allow a user to search the entire Internet for key words or phrases

in return, charges a monthly fee. Once the Internet has been accessed, **search engines** can be used to search for documents or web pages that would have the words or phrases being searched for (see Figure 12-4). For example, suppose a safety program manager wanted to obtain more information about exposures to bloodborne pathogens. Using the search feature of the Internet and typing in the

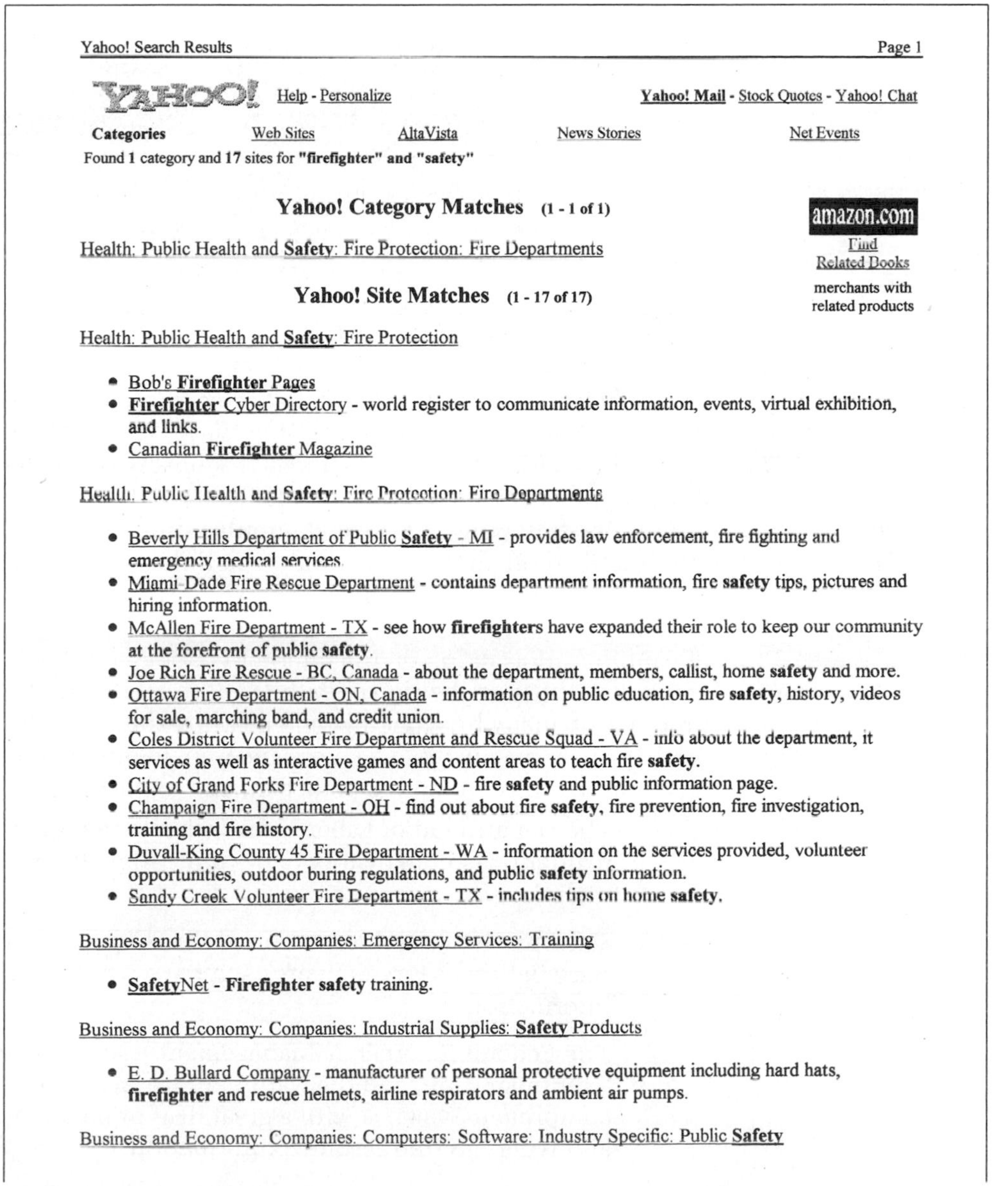

Yahoo! Search Results Page 1

YAHOO! Help - Personalize **Yahoo! Mail** - Stock Quotes - Yahoo! Chat

Categories Web Sites AltaVista News Stories Net Events

Found 1 category and 17 sites for **"firefighter" and "safety"**

Yahoo! Category Matches (1 - 1 of 1)

amazon.com
Find Related Books
merchants with related products

Health: Public Health and **Safety**: Fire Protection: Fire Departments

Yahoo! Site Matches (1 - 17 of 17)

Health: Public Health and **Safety**: Fire Protection

- Bob's **Firefighter** Pages
- **Firefighter** Cyber Directory - world register to communicate information, events, virtual exhibition, and links.
- Canadian **Firefighter** Magazine

Health: Public Health and **Safety**: Fire Protection: Fire Departments

- Beverly Hills Department of Public **Safety** - MI - provides law enforcement, fire fighting and emergency medical services.
- Miami-Dade Fire Rescue Department - contains department information, fire **safety** tips, pictures and hiring information.
- McAllen Fire Department - TX - see how **firefighters** have expanded their role to keep our community at the forefront of public **safety**.
- Joe Rich Fire Rescue - BC, Canada - about the department, members, callist, home **safety** and more.
- Ottawa Fire Department - ON, Canada - information on public education, fire **safety**, history, videos for sale, marching band, and credit union.
- Coles District Volunteer Fire Department and Rescue Squad - VA - info about the department, it services as well as interactive games and content areas to teach fire **safety**.
- City of Grand Forks Fire Department - ND - fire **safety** and public information page.
- Champaign Fire Department - OH - find out about fire **safety**, fire prevention, fire investigation, training and fire history.
- Duvall-King County 45 Fire Department - WA - information on the services provided, volunteer opportunities, outdoor buring regulations, and public **safety** information.
- Sandy Creek Volunteer Fire Department - TX - includes tips on home **safety**.

Business and Economy: Companies: Emergency Services: Training

- **Safety**Net - **Firefighter safety** training.

Business and Economy: Companies: Industrial Supplies: **Safety** Products

- E. D. Bullard Company - manufacturer of personal protective equipment including hard hats, **firefighter** and rescue helmets, airline respirators and ambient air pumps.

Business and Economy: Companies: Computers: Software: Industry Specific: Public **Safety**

Figure 12-4 *The first page of results from an Internet search on "safety" and "firefighter."*

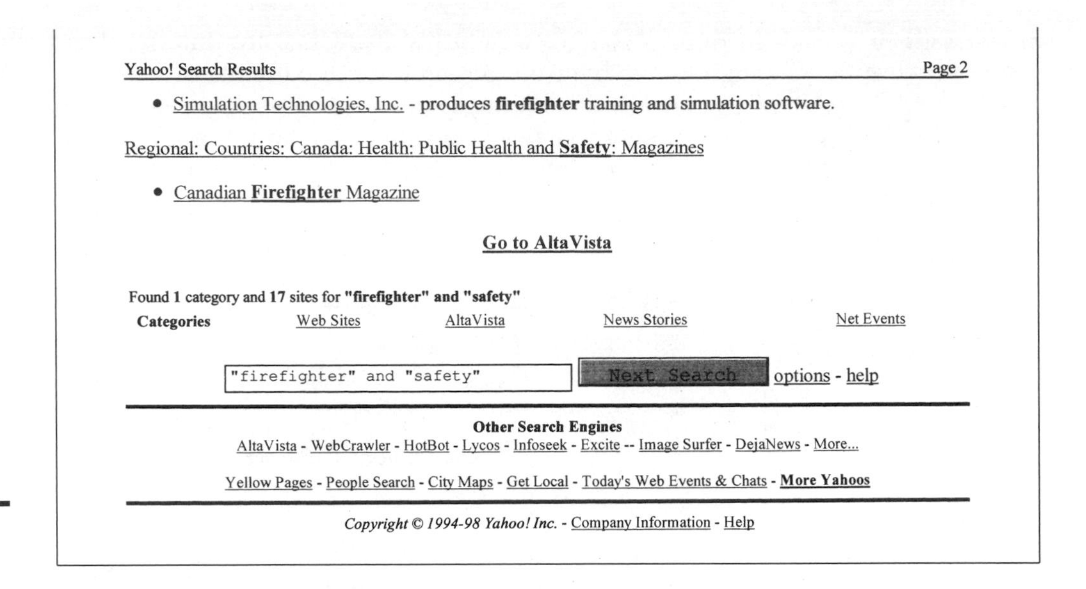
Yahoo! Search Results Page 2

- Simulation Technologies, Inc. - produces **firefighter** training and simulation software.

Regional: Countries: Canada: Health: Public Health and **Safety**: Magazines

- Canadian **Firefighter** Magazine

Go to AltaVista

Found 1 category and 17 sites for **"firefighter" and "safety"**

Categories Web Sites AltaVista News Stories Net Events

"firefighter" and "safety" Next Search options - help

Other Search Engines

AltaVista - WebCrawler - HotBot - Lycos - Infoseek - Excite -- Image Surfer - DejaNews - More...

Yellow Pages - People Search - City Maps - Get Local - Today's Web Events & Chats - **More Yahoos**

Copyright © 1994-98 Yahoo! Inc. - Company Information - Help

Figure 12-4
Continued

hits
number of documents found when searching for a key word or phrase using a search engine

links
used on Internet pages so that each page may be directly tied to a page on another Internet site

words *bloodborne* and *pathogens,* the search engine searches the Internet for documents and web pages containing these words. The search engine then displays a list of **hits.** The user can then click on any of the documents on the list and the Internet will connect them to that site.

In the following list, several popular sites for safety and health information are described. Each also commonly has **links** to other pages with similar information. The reader is encouraged to visit these sites and to search for others. The information gained is timely and useful to the safety program. However, a note of caution: Information on the Internet is not guaranteed to be factual. If there is any doubt, the user should verify the validity of the information. Often the search produces so many hits that it would be unreasonable to look at all of them. In this case, the user may want to be more specific. In our example, the next search may have the words bloodborne, pathogen, and exposure.

The Department of Labor has a web site that gives access to OSHA regulations and a history of organizations that were cited for specific violations.

The International Association of Fire Chiefs and the International Association of Fire Fighters each have web sites that can provide usable information.

The Federal Emergency Management Agency, United States Fire Administration, and the National Fire Academy have very comprehensive sites with a great deal of health and safety information and where to find resource information.

The Centers for Disease Control site can be helpful in obtaining information regarding infectious diseases and information on how to comply with regulations dealing with infectious diseases.

At least one workers' compensation carrier is allowing for Internet access to an organization's claim records. The access is limited to those who need the information for tracking claims or who are responsible for the health and safety program.

A book could be written on the Internet and the relation to emergency services. Instead, Appendix II lists sites pertinent to the health and safety of emergency responders. This list is not exhaustive but a good start.

Summary

Data collection and information analysis are important components of the safety and health program. The data collection needs will vary based on whether the information is for use within the department or for an external agency. Each can be turned into valuable feedback that will assist in the evaluation of the program. Regardless of where the information is to be used, a form or similar consistent means should be established for gathering the data.

Once analyzed, the information should be compiled into an annual report for distribution organizationwide. The annual report has a number of required sections and should answer the questions of where were we, where are we now, where are we going, and how can we get there?

By using computers and the Internet, an organization can communicate much faster and can access much more information more easily than has ever been possible. Many national, state, and local health and safety-related agencies have information on the Internet, as do local emergency response organizations. This information can be a valuable tool in assisting the health and safety program manager and safety committee in program design, training, and evaluation.

Concluding Thought: As with the other components of the program, information management is essential. Through the use of computers and the Internet, there is virtually no limit to the access of this information and new information.

Review Questions

1. List three reasons for data collection and reporting.
2. Compare date collection from the internal perspective to the external perspective.
3. Injury data downloaded from the USFA web site would be considered internal data.
   A. True
   B. False
4. Which of the following are essential to an organization's information system?
   A. Data must be collected using a common means.
   B. Data must be stored so that it is easily retrievable.
   C. Data must be analyzed and put in a useful format.
   D. All of the above are essential components.
5. List the sections that should be a part of the annual safety and health report.

Activities

1. Compare the annual health and safety report from your department with the recommendations in this text. Are all the sections included? Is the document used as a planning tool?
2. Analyze the procedures in your department for collecting and storing health and safety data. Are the data retrievable in a usable format? Are the data useful as information?
3. Access the Internet if possible. Visit the sites that are recommended in the text and search for others. What information can be useful in your department?

Emerging Issues and Trends

Learning Objectives

Upon completion of this chapter, you should be able to:

- Recognize controversial issues as they relate to the safety and health program.
- Describe health and safety considerations in the diversified workforce.
- Describe the health and safety program implications of future technologies.

CASE REVIEW

It is a cold, snowy night in the year 2025 in a small, northeastern town on the Atlantic Ocean. The family retires to bed around 11:00 P.M. Of course, the heating system in the house is working overtime, as the temperatures in the region have been below freezing for days. Around 1:00 A.M. as the four members of the family sleep, an electrical circuit in the wall begins to overheat. Although the computer-based home warning system detects the overheating, the system malfunctions and the hot wire starts a fire in the concealed spaces of the wall. As the fire grows, slowly a sensor mounted in the ductwork detects the presence of a flame and sounds the interior fire alarm and lights the pathway to the exits. Had the fire been outside the wall, the quick response sprinkler heads probably would have activated. The family awakes to a light haze in the house and retreats to the prearranged meeting place outside of the house. They hear sirens approaching, as the fire department was notified when the flame sensor in the wall activated. The response to the alarm was quick, because the fire department had the on-board computer and global positioning system (GPS), so the maps lead the responders directly to the incident location. Of course, using the GPS as a locator allowed the fire officer to access a floor plan of the house and any special hazards, were noted. Upon arrival, the firefighters entered the house with the latest in personal protective equipment. The incident commander (IC), using information transmitted from the house, can see exactly where the fire is located. The IC also knows that rescue is not a problem as she has spoken to the occupants. The IC watches the movements of all personnel on the scene on the computer monitor. The status of the firefighters is also displayed, including body temperature, hydration level, amount of air left, blood pressure and pulse rates, and the status of the personal alert safety system (PASS) device.

Firefighters can see very well in the home as a result of their thermal imaging devices built into their self-contained breathing apparatus (SCBA) masks. The heads up display on the masks provides a floor plan and essential information about the extent of the fire and the status of their protective equipment. As the IC monitors the situation from the command vehicle, the firefighters get to the exact location of the fire and, using sound waves through the wall, extinguish the fire. Roof vents are automatically operated from the incident commander's computer and the built-in ventilation system comes on. An air freshener is introduced to limit the odor of smoke in furnishings in the house. The entire process takes less than 10 minutes. Of course, the firefighters could have worked a lot longer, as they are in top physical condition and air supply is no longer a problem, since cryogenic-based liquid air systems were introduced in 2000.

The incident commander chats with the family and instructs them to follow up in the morning with an electrician. The fire units go back into service immediately, because no equipment had been removed from the apparatus and no hose had to be reloaded. Of course, there were no injuries; they had not had an injury since 2001.

INTRODUCTION

This text would not be complete without some discussion of the controversial issues affecting present and future safety and health programs. These areas, although applicable to the many subjects already presented, fit better into this chapter because they are issues that do not specifically fall within the preceding chapters. Much of the information in the remainder of this chapter has come from opinions and from various sources. It should not be considered the last word on a subject, but instead is designed to spark interest, future research, and a scanning of changes in the emergency services field in order to be prepared for the challenges that are sure to lie ahead in the twenty-first century.

CONTROVERSIAL ISSUES

Have we gone too far or not far enough? Difficult question. The answer depends on who you talk to. Clearly, by virtue of writing this book, I believe that more can be done. However, some believe that the safety pendulum has swung too far to the pro-safety side. As a result, they say emergency service providers, by taking a more cautious systematic approach, are doing a disservice to the citizens and that we are forgetting our commitment to public safety above all else. The pro-safety people believe that the responders' safety is equally as, if not more, important than the safety of the person or property we are called upon to protect.

In Table 13-1, I have listed some issues that are controversial, based on numerous discussions, magazine articles, seminars, and other educational set-

Table 13-1 *Controversial issues.*

Issue	Are we doing enough?	Have we gone too far?
Hazardous material response	Way too much emphasis on full protection when often the quantities or composition of the product do not warrant such action.	Full protection is the only way to ensure safety and no exposure to the product.
Firefighting protective clothing	Too much of an envelope; cannot feel the heat burning the ears; will not know when to get out.	Full protection against flashover. Knowing when to get out should be a part of training, understanding fire behavior, and good tactics.
Bunker pants vs. 3/4 boots	Bunker pants too hot, will cause heat exhaustion; limit motion, hard to climb stairs.	Provide greater level of protection. Heat exhaustion handled through proper rehabilitation.
Critical incident stress management	We never did that before. We handled the stress in our own ways.	Needed to provide an avenue of stress release and to ensure maximum mental preparedness of responders.
Incident management system	Spending too much time setting up command systems; no one is taking care of the incident.	Must begin at the arrival of the first or second unit to ensure a continuity throughout the incident.
Accountability	Difficult to manage; why do it on single alarm incidents?	Must start at the beginning of the incident with the arrival of the first units. Many system models are not difficult to work with.
Offensive fire attack on vacant buildings	The perception would be that the fire department is not doing what they are expected to do; no building is vacant until searched.	Risk benefit approach; if there is not a reason to believe that a life is at risk, do not put a responder's life at risk.

tings. As stated in the introduction to this chapter, I hope they will spark interest and debate. The comments as written are at opposite ends of the spectrum, the best position might be somewhere in the middle.

SAFETY CONSIDERATIONS IN THE DIVERSIFIED WORKFORCE

According to a report called *Workforce 2000*,[1] the diversity in the workplace will continue to increase. While this document was written for the workforce as a whole, there are implications for the emergency service occupations. Some of the findings from the report include:

- Population growth will slow.
- Average age of the workforce will rise.
- More woman will enter the workforce.
- White males will be a smaller percentage of the workforce.
- Minorities will be a larger share of the new entrants into the labor force.

Although there are no health and safety considerations for all of these findings, a couple do support some discussion. The increasing average age of the workforce will require continued emphasis on physical fitness. Although the emergency services have generally adapted well to women in the workforce, their entry has been relatively recent. What safety and health considerations must be undertaken for women who are pregnant? Policies must be developed before the situation arises. Many chemicals can affect the fetus, as well as the mother.

The design and location of response apparatus must be considered as well. Part of a diverse workforce is variations in height. Years ago, fire departments had height requirements. After these were challenged and removed, there became a need for a redesign of apparatus, particularly ladder storage, patient compartment height, and the height of hose beds.

FUTURE TECHNOLOGIES AND TRENDS

Probably one of the more exciting prospects in the field of health and safety is looking toward the future. Just consider for a moment the technological changes that have occurred over the last 20 years. Look at the protective equipment changes and how the emergency services have had to adapt. Many items are coming out now that are somewhat expensive and out of reach of some departments, but as technology improves, the market will drive the price down. Remember the first personal computers? Compare their price and capabilities to what you can buy today for a fraction of the cost.

---

[1] Published in 1987 by the Hudson Institute of Indianapolis for the Department of Labor.

The response unit that we use will also change. Major emergency vehicle manufacturers have already unveiled prototype apparatus of the future. These apparatus will be lighter, smaller, and multipurpose. Responder's safety issues, such as equipment storage and access, height of components, and the safety of the crew compartment, will all be considerations in the design (see Figure 13-1). The following is a list of some of the things that probably will be as common as the fire helmet in 5 to 10 years.

- Emergency Scene GPS linked to the incident commander's computer for location of all personnel.
- Individual status information projected on the SCBA mask.
- The ability to have building floor plans projected on the SCBA facemask.
- Better early warning for fires and better private fire protection.
- Greater use of GPS and intersection control to reduce intersection accidents.
- Greater use of thermal imaging systems (see Figure 13-2).
- Further improvements in protective clothing to prevent exposures.
- Change from compressed air to cryogenic air supply, reducing the weight and increase the duration of SCBA.
- Greater emphasis on occupational safety—hiring specific health and safety officers with higher-level degrees in occupational safety and health.
- Greater emphasis on research for health and safety programs including a study of what is done in other countries, giving health and safety solutions a global flavor.
- Better training aids or realistic training (see Figure 13-3).

Figure 13-1 *Fire apparatus design will change in the future, with responder safety playing a role. (Courtesy of E-One, Inc., Ocala, FL.)*

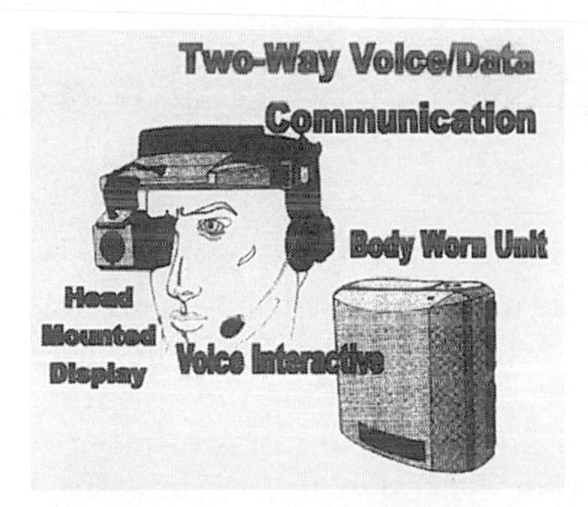

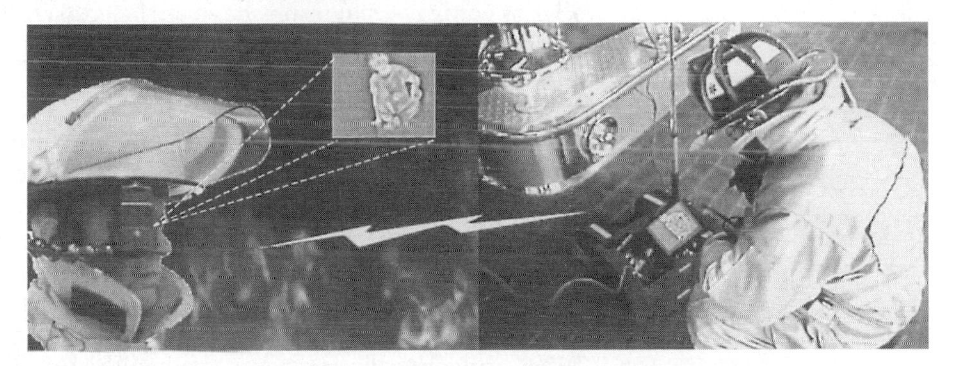

Figure 13-2 *Firefighters in the future will have a great deal more information available. (Courtesy of Naval Air Warfare Center, Aircraft Div., Willow Grove, PA.)*

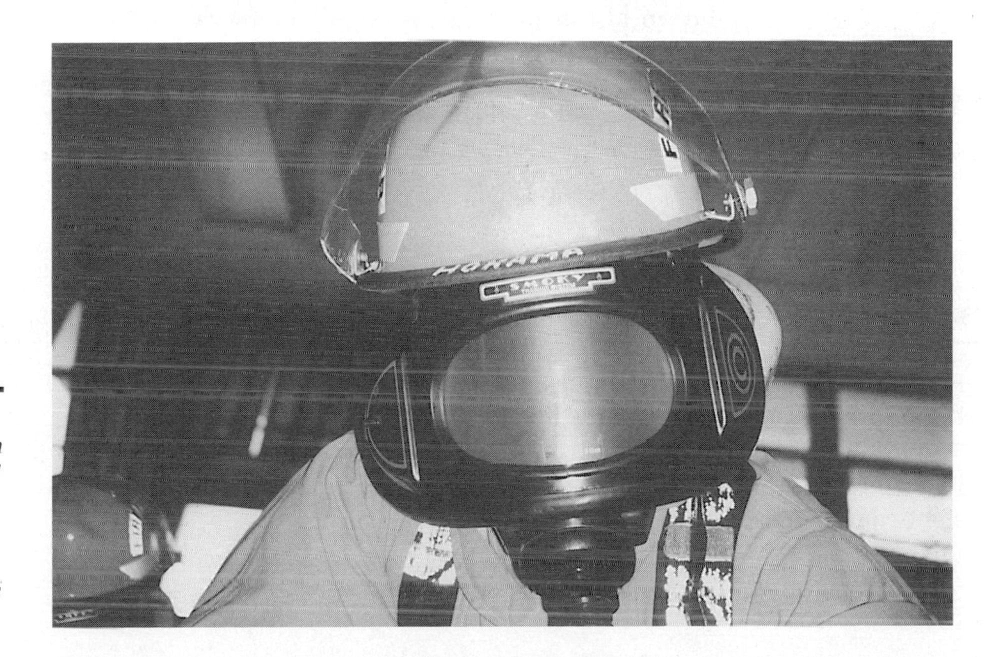

Figure 13-3 *This training aid clips on the SCBA mask and simulates various levels of reduced visibility. (Courtesy of Robwen, Inc., Los Angeles, CA.)*

Summary

Although the emergency service responders and organizations have made significant improvements in the attention given to health and safety, some say that the process has gone too far, while others say that more can be done. There is not a clear answer to this debate. However, when examining the statistics it would appear clear that there are still unreasonable numbers of emergency service injuries and deaths each year. Through the various approaches to health and safety, a number of controversial issues have surfaced. While the arguments have validity depending on what side you are on, there is probably a middle ground in which both sides can agree.

Changes in the diversity and demographics will have implications for the health and safety program manager and to the overall program. These changes have been identified in the workforce 2000 document and focus on a slowing population growth, a rise in the average age on the workforce, more woman entering the workforce, white males becoming a smaller percentage of the workforce, and minorities becoming a larger share of the new entrants into the workforce. There are health and safety considerations in several of these areas.

Finally, future technology and trends will affect the safety and health program. From better protective equipment to the health and safety manager as a professional similar to a fire protection engineer, the focus will be on continued support and improvement of the safety and health for the emergency responders of the world.

Concluding Thought: The only thing certain is that things will change. Many changes and future technologies will focus on responders' safety. We must be ready to accept these changes and use them to the advantage of the safety and health program.

Review Questions

1. List the predictions of workforce 2000, and identify those that will have a health and safety impact.
2. List three controversial issues related to safety and health.
3. List five implications of future technology to a health and safety program.
4. List two trends that will affect the health and safety program.
5. What are the implications of an older workforce to the health and safety program?

Activities

1. Examine the workforce 2000 predictions. How will they affect your department?
2. Pick any two of the controversial issues, and write a position on them, either pro or con.
3. Pick any two future technologies or trends, and discuss their implications in your department.

Reprint of the 1996 Firefighter Fatality Survey from the United States Fire Administration

Although this text only includes the report from 1996, the reports are done annually. The report is included to introduce the reader to the information available from a report of this nature. The reader is encouraged to obtain a current report from the publications center of the USFA or from the USFA Internet site for local comparison.

FIREFIGHTER FATALITIES IN THE UNITED STATES IN 1996

Prepared for

United States Fire Administration
Federal Emergency Management Agency
Contract No. EMW-95-C-4713

Prepared by

TriData Corporation
1000 Wilson Boulevard
Arlington, Virginia 22209

August 1997

TABLE OF CONTENTS

ACKNOWLEDGMENTS

This study of firefighter fatalities would not have been possible without the cooperation and assistance of many members of the fire service across the United States. Members of individual fire departments, chief fire officers, the National Interagency Fire Center, US Forest Service personnel, the US military, the Department of Justice, and many others contributed important information for this report.

TriData Corporation of Arlington, Virginia, conducted this analysis, for the United States Fire Administration under contract EMW-95-C-4713.

The ultimate objective of this effort is to reduce the number of firefighter deaths through an increasing awareness and understanding of their causes and how they can be prevented. Firefighting, rescue, and other types of emergency operations are essential activities in an inherently dangerous profession, and tragedies will occur from time to time. This is the risk all firefighters accept every time they respond to an emergency incident. However, the risk can be greatly reduced through efforts to increase firefighter health and safety.

The United States Fire Administration would like to extend its thanks to the Boston Fire Department's Public Relations Office for providing the photograph for the cover. The picture was taken during the dedication ceremony for the Vendome Memorial in Boston, MA. The Hotel Vendome fire occurred on June 17, 1972, and killed nine Boston firefighters.

This report is dedicated to those firefighters who have made the ultimate sacrifice in 1996. May the lessons learned from their passing not go unheeded.

BACKGROUND

For the last 20 years, the United States Fire Administration (USFA) has tracked the number of firefighter fatalities and conducted an annual analysis. Through the collection of information on the causes of firefighter deaths, the USFA is able to focus on specific problems and direct efforts towards finding solutions to reduce the number of firefighter fatalities in the future. This information is also used to measure the effectiveness of current programs directed toward firefighter health and safety.

In addition to the analysis, the USFA maintains a list of firefighter fatalities for the Fallen Firefighter Memorial Service. The fallen firefighters' next of kin as well as members of the individual fire departments are invited to the annual Fallen Firefighter Memorial Service, which is held at the National Fire Academy in Emmitsburg, Maryland, every fall. Additional information regarding the memorial service can be found on the Internet at http\\www.usfa.fema.gov or by calling the National Fallen Firefighters Foundation at 301-447-1365.

INTRODUCTION

This report continues a series of annual studies by the US Fire Administration of firefighter fatalities in the United States.

The specific objective of this study was to identify all of the on-duty firefighter fatalities that occurred in the United States in 1996, and to analyze the circumstances surrounding each occurrence. The study is intended to help identify approaches that could reduce the number of firefighter deaths in future years.

In addition to the 1996 overall findings, this study includes special analyses on violent firefighter deaths, physical fitness and its relation to firefighter deaths, and vehicle accidents.

WHO IS A FIREFIGHTER?

For the purpose of this study, the term *firefighter* covers all members of organized fire department, including career and volunteer firefighters; full-time public safety officers acting as firefighters; state and federal government fire service personnel, including wildland firefighters; and privately employed firefighters, including employees of contract fire departments and trained members of industrial fire brigades, whether full or part-time. It also includes contract personnel working as firefighters or assigned to work in direct support of fire service organizations.

Under this definition, the study includes not only local and municipal firefighters, but also seasonal and full-time employees of the United States Forest Service, the Bureau of Land Management, the Bureau of Indian Affairs, the Bureau of Fish and Wildlife, the National Park Service, and state wildland agencies. It also includes prison inmates serving on firefighting crews; firefighters employed by other governmental agencies such as the United States Department of Energy; military personnel performing assigned fire suppression activities; and civilian firefighters working at military installations.

WHAT CONSTITUTES AN ON-DUTY FATALITY?

On-duty fatalities include any injury or illness sustained while on-duty that proves fatal. The term *on-duty* refers to being involved in operations at the scene of an emergency, whether it is a fire or non-fire incident; being en route to or returning from an incident; performing other officially assigned duties such as training, maintenance, public education, inspection, investigations, court testimony and fund-raising; and being on-call, under orders, or on stand-by duty, except at the individual's home or place of business.

These fatalities may occur on the fireground, in training, while responding to or returning from alarms, or while performing other duties that support fire service operations.

A fatality may be caused directly by accident or injury, or it may be attributed to an occupational-related fatal illness. A common example of a fatal illness incurred on duty is a heart attack. Fatalities attributed to occupational illnesses would also include a communicable disease contracted while on duty that proved fatal, where the disease could be attributed to a documented occupational exposure.

Accidents that claim the lives of on-duty firefighters are also included in the analysis, whether or not they are directly related to emergency incidents. In 1996, this category includes a firefighter who died in a car accident while in transit between fire station assignments and a total of six firefighters who were victims of violence against emergency service personnel.

Injuries and illnesses are included where death is considerably delayed after the original incident. When the incident and the death occur in different years, the analysis counts the fatality as having occurred in the year that the incident occurred. For example, a firefighter died in 1996 of medical complications that resulted from an injury at a fire in 1982. Because his death was the result of the 1982 incident, this case was counted as a 1982 fatality for statistical purposes, and is not included in the 94 fatalities for 1996 that were analyzed in this report. Since the death occurred in 1996, he will be included in the 1996 annual Fallen Firefighter Memorial Service at the National Fire Academy, and his name will be included on the list of firefighters who died in 1996.

There is no established mechanism for identifying fatalities that result from illnesses that develop over long periods of time, such as cancer, which may be related to occupational exposure to hazardous materials or products of combustion. It has proven to be very difficult over several years to provide a full evaluation of an occupational illness as a causal factor in firefighter deaths, because of the limitations in the ability to track the exposure of firefighters to toxic hazards, the often delayed long-term effects of such exposures, and the exposures firefighters may receive while off-duty.

SOURCES OF INITIAL NOTIFICATION

As an integral part of its ongoing program to collect and analyze fire data, the United States Fire Administration solicits information on firefighter fatalities directly from the fire service and from a wide range of other sources. These include the Public Safety Officer's Benefit Program (PSOB) administered by the Department of Justice, the Occupational Safety and Health Administration (OSHA), the US military, the National Interagency Fire Center, and other federal agencies.

The USFA receives notification of some deaths directly from fire departments, as well as from fire service organizations such as the International Association of Fire Chiefs (IAFC), the International Association of Fire Fighters (IAFF), the National Fire Protection Association (NFPA), the National Volunteer Fire Council (NVFC), state fire marshals, state training organizations, other state and local organizations, and fire service publications. The USFA also keeps track of fatal fire incidents as part of its Major Fire Investigations Project and maintains an ongoing analysis of data from the National Fire Incident Reporting System (NFIRS) for the production of the report *Fire in the United States*.

PROCEDURE FOR INCLUDING A FATALITY IN THE STUDY

In most cases, after notification of a fatal incident, initial telephone contact is made with local authorities by the USFA's contractor to verify the incident, its location and jurisdiction, and the fire department or agency involved. Further information about the deceased firefighter and the incident may be obtained from the chief of the fire department or his designee over the phone or by other data collection forms.

Information that is routinely requested includes NFIRS-1 (incident) and NFIRS-3 (fire service casualty) reports, the fire department's own incident reports and internal investigation reports, copies of death certificates or autopsy results, special investigative reports such as those produced by the USFA or NFPA, police reports, photographs and diagrams, and newspaper or media accounts of the incident.

After obtaining this information, a determination is made as to whether the death qualifies as an on-duty firefighter fatality according to the previously described criteria. The same criteria was used for this study as in previous annual studies. Additional information may be requested, either by follow-up with the fire department directly, or from state vital records offices or other agencies. The determination as to whether a fatality qualifies as an on-duty death for inclusion in the statistical analysis and the Fallen Firefighter Memorial Service is made by the USFA.

1996 FINDINGS

Ninety-four (94) firefighters died while on duty in 1996.[1] This is a slight decrease from last year's total of 96. The total of 94 fatalities is the third lowest number recorded in the 20 years that this data has been collected, and is only the fourth time that the total has been less than 100 fatalities. The lowest years were 1992 with 75 fatalities and 1993 with 77 fatalities.

This year's total is part of a long-term downward trend of reduced fatalities that began in 1979 after a peak of 171 in 1978. The overall trend in firefighter fatalities is down 35 percent over the last ten years. Over the last five years there has been an upward trend of 29 percent, though the number of deaths in 1996 decreased approximately two percent from 1995 (Figure 1).

Figure 1 On-Duty Firefighter Deaths for 1996

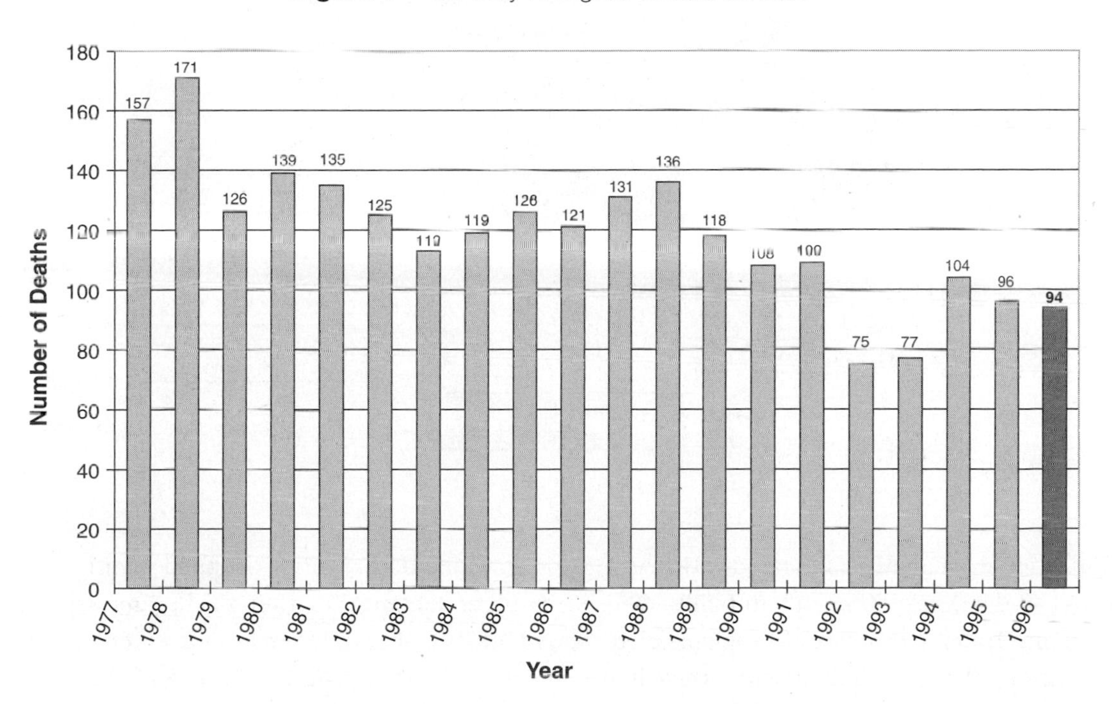

[1] As mentioned earlier, the 94 on-duty fatalities in 1996 do not include one firefighter who died during the year from injuries sustained in 1982. This firefighter was injured when a concrete loading dock collapsed at a paper warehouse fire. He was in a coma for 13 years before he died.

The fatalities included 68 volunteer firefighters and 26 career firefighters (down from 33 career in 1994) (Figure 2). Among the volunteer firefighter fatalities, 62 were from local or municipal volunteer fire departments, 3 were part-time or seasonal members of wildland fire agencies, and 3 were members of department fire-police units. All the career firefighters who died were members of local or municipal fire departments. Ninety-one of the fatalities were men and three were women.

The 94 deaths resulted from 89 incidents. Three multi-fatality incidents resulted in 8 firefighter deaths. Four firefighters died in Jackson, Mississippi when a disgruntled firefighter shot coworkers in an administrative meeting. Two firefighters were killed when they lost control of their vehicle, ran off the road, overturned, and hit a tree (the call turned out to be a false alarm). Another two firefighters died in a commercial structure fire when a lightweight truss roof collapsed.[2]

Figure 2 Career vs. Volunteer Deaths

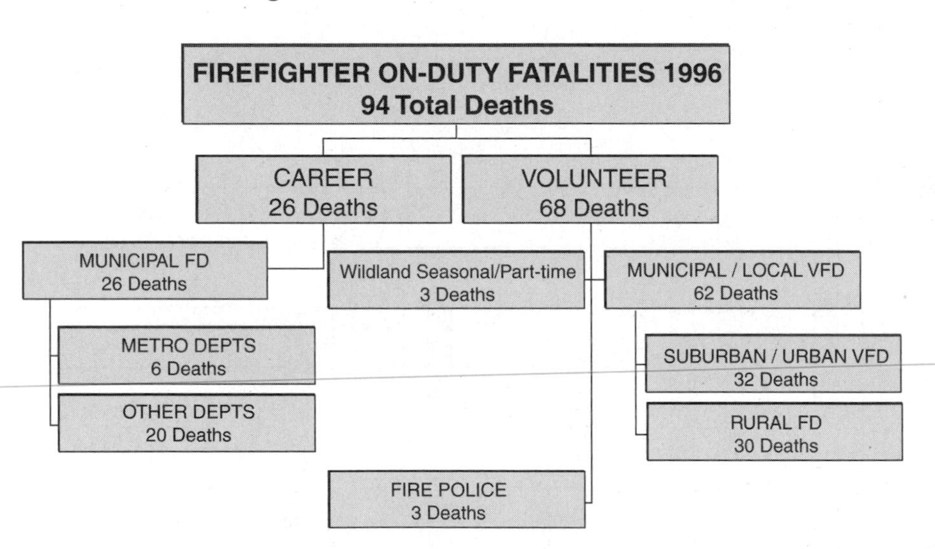

The number of deaths associated with brush, grass or wildland firefighting dropped significantly from 18 in 1995 to five deaths in 1996. One firefighter died as a result of dehydration when she became separated from her group on a training run. Four firefighters died of heart attacks, one while repairing a water tender between fires, one after fighting a wildland fire for five hours, one while fighting a wildland fire, and one during a tree-felling class.

[2] The report "Two Firefighters Killed in Chesapeake, VA" can be ordered from the USFA.

TYPE OF DUTY

In 1996, 68 firefighter on-duty deaths were associated with emergency incidents, accounting for 72 percent of the 94 fatalities (Figure 3). This includes all firefighters who died while responding to an emergency, while at the emergency scene, or after the emergency incident. Non-emergency activities accounted for 26 fatalities (28 percent). Non-emergency duties include training, administrative activities, or performing other functions that are not related to an emergency incident. One firefighter working at a department fund-raiser and another firefighter directing parade traffic were included in this number.

Figure 3 Firefighter Deaths While Performing Emergency Duty 1996

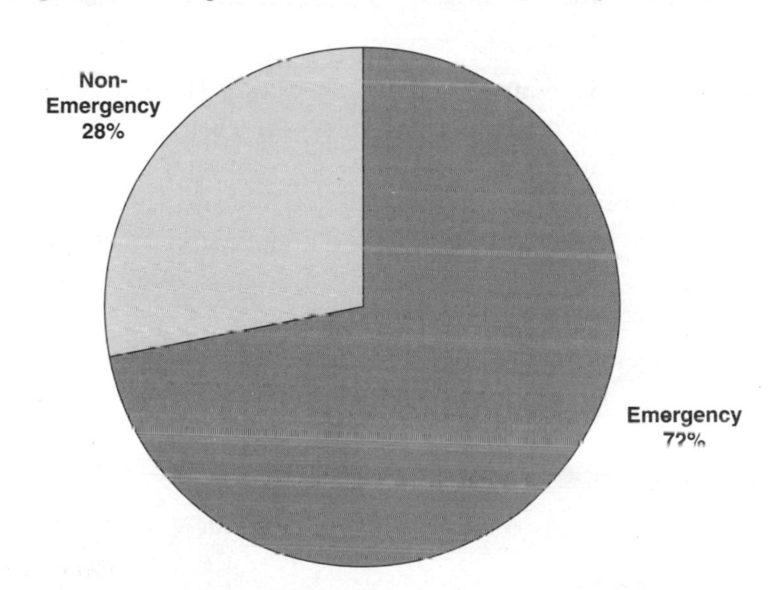

The number of deaths by type of duty being performed is shown in Table 1 and presented graphically in Figure 4. As in previous years, the largest number of deaths occurred during fireground operations. There were 38 fireground deaths, which accounted for 40 percent of the fatalities, down two percent from 1995. Over half (23) resulted from heart attacks on the scene. Eight were from asphyxiation, three from internal trauma, two from electrocution, one from a pulmonary edema, and one from burn injuries.

The second largest category of deaths by duty type was responding to or returning from emergency incidents, which accounted for 22 deaths in 1996 (down nine deaths from 1995). This has been the second leading cause of deaths since 1993. All 22 deaths involved volunteer firefighters. Six firefighters suffered fatal heart attacks while responding to or returning from emergency incidents. Eight firefighters were killed in fire apparatus accidents while enroute to emergency incidents. At least five of these deaths involved apparatus rollovers.

Table 1 Type of Duty–1996

	Number	Percent
Fireground Operations	38	40.4%
Responding/Returning from Alarm	22	23.4%
Other/On-Duty	20	21.3%
Non-Fire Emergencies	8	8.5%
Training	6	6.4%
TOTAL	94	100%

Eight firefighters were killed in accidents involving their personal vehicles while enroute to emergency calls. One of these involved a firefighter who drowned in a roadside lake when his personal vehicle wrecked on the way to a call.

Eight deaths were related to activities at the scene of non-fire emergency incidents. This is down from 13 deaths in 1995. Three firefighters died of heart attacks during EMS incidents. One firefighter was killed

Figure 4 Type of Duty–1996

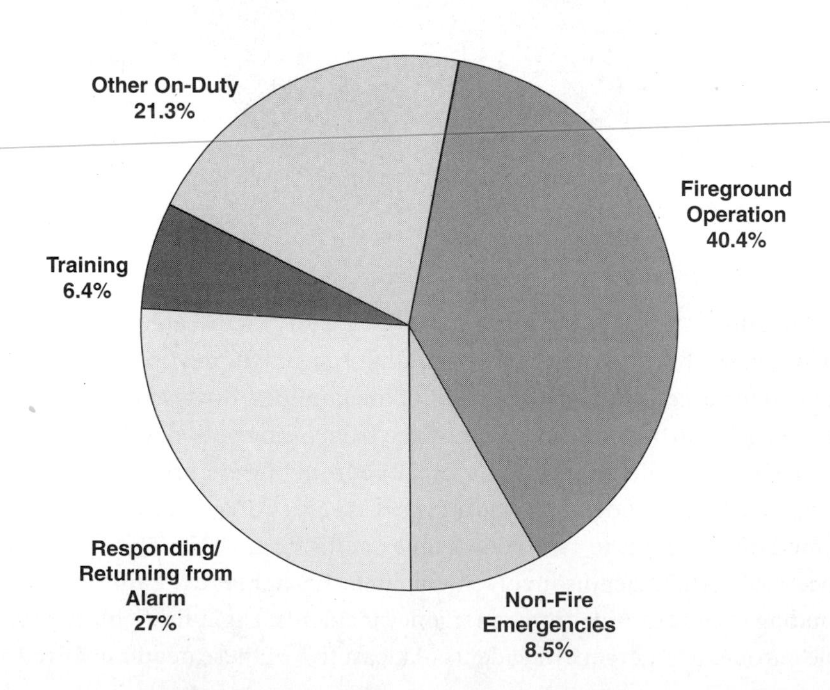

when he was electrocuted at a motor vehicle accident. Another firefighter died of asphyxiation during a technical rescue in a grain bin. One firefighter was killed when he was shot by an irate victim at the scene of a motor vehicle accident (MVA), and another firefighter died due to complications from an injury sustained at an MVA that occurred while transporting a patient to the hospital. Another firefighter died when he was hit by a passing motorist while extricating a patient from a vehicle.

There were 26 deaths that occurred during non-emergency duty activities. These deaths include nine firefighters who died from heart attacks while on duty—two at fire department fund-raisers, two during the night at the fire department, two while exercising, one while performing a stress and agility test, one while repairing apparatus between calls, and one while directing traffic for a parade as part of his fire police duties. One firefighter died of a stroke while inspecting fire hydrants. Ten of the 26 non-emergency duty deaths were a result of internal trauma—four firefighters were shot and killed by a disgruntled firefighter during an administrative meeting at the fire department, one firefighter was stabbed to death on the way to storm duty, two firefighters died in motor vehicle accidents (one in transit between station assignments and another while returning from a non-emergency service call), one firefighter was struck by a vehicle while directing traffic, one firefighter fell 20 feet down a fire pole hole in the station, and one was killed during a Fourth of July celebration sponsored by the fire department. The latter was a licensed pyrotechnician who was killed when fireworks prematurely detonated.

Six deaths were attributed to training activities, including one death from dehydration when the firefighter became separated from the group during a training run. Four firefighters died as a result of heart attacks during training—one at a live burn training session, one after an EMS training event, one during a vehicle extrication drill, and one during a wildland tree-felling class. One firefighter died during recruit training from massive organ failure due to a pre-existing condition of sickle cell anemia.

CAUSE OF FATAL INJURY

As used in this study, the term *cause* of injury refers to the action, lack of action, or circumstances that directly resulted in the fatal injury, while the term *nature* of injury refers to the medical cause of the fatal injury or illness, often referred to as the physiological cause of death. A fatal injury usually is the result of a chain of events, the first of which is recorded as the cause. For example, if a firefighter is struck by a collapsing wall, becomes trapped in the debris, runs out of air before being rescued, and dies of asphyxiation, the cause of the fatal injury is recorded as "struck by collapsing wall" and the nature of the fatal injury is "asphyxiation". Similarly, if a wildland firefighter is overrun by a fire and dies of burn, the cause of the death would be listed as "caught/trapped," and the nature of death would be "burns". This follows the convention used in NFIRS casualty reports.

Figure 5 shows the distribution of deaths by cause of fatal injury or illness and Table 2 presents the exact number. As in most previous years, the largest cause category is stress or overexertion, which was listed as the primary factor in 50 percent of the deaths, the same as last year. Firefighting has been shown to be one of the most physically demanding activities that the human body performs. Most firefighter deaths attributed to stress result from heart attacks. Of the 47 stress-related fatalities in 1996, 44 firefighters died of heart attacks, one

Figure 5 Cause of Fatal Injury

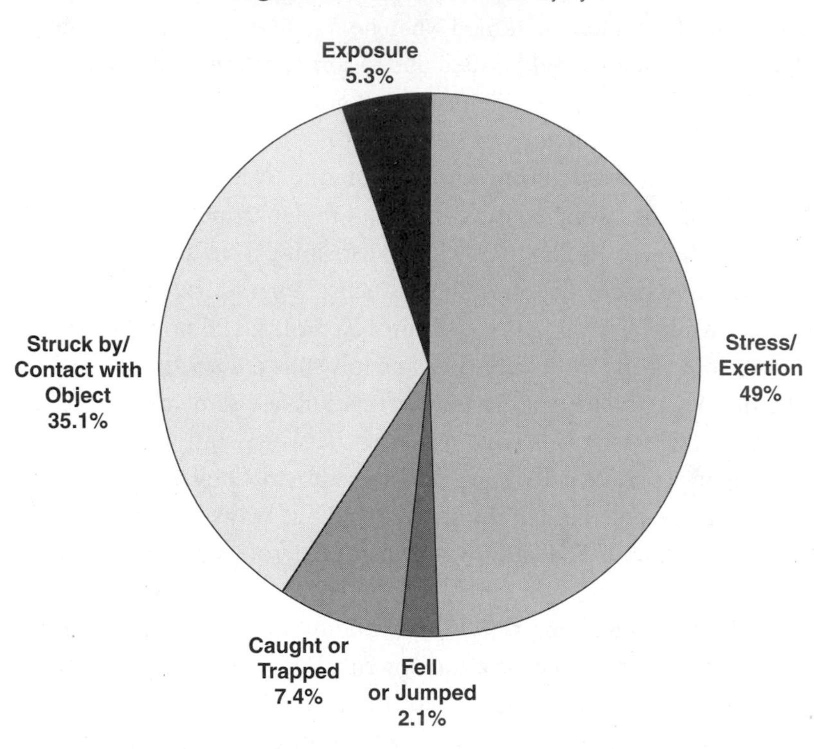

died of a stroke, one died from complications with sickle cell anemia[3], and one died of a dehydration. Sixteen of the 47 deaths whose cause is listed as stress/exertion occurred during non-emergency activities.

The second leading cause of firefighter fatalities was being struck by or coming in contact with an object. Of the 33 firefighters (35 percent) who died in these incidents, 17 were involved in vehicle accidents, six were murdered, four were struck by vehicles, two were electrocuted, two were killed when their truck was struck by a falling tree, one was struck by detonated fireworks, and one was killed by a collapsing wall.

The third leading cause of firefighter fatalities was being caught or trapped, which accounted for seven deaths (7.4 percent), down 14 percent from 1995. Six firefighters died after becoming trapped by roof collapses—five were in commercial structures and one in a townhouse. One firefighter drowned after being caught in his car after running off the road into a lake.

Three asphyxiation deaths were attributed to exposure[4]. One was a firefighter who died when he took off his SCBA (self-contained breathing apparatus) mask in an oxygen deficient atmosphere. One firefighter

---

[3] Due to the pre-existing condition of sickle cell anemia, the firefighter after intense exertion went into a state of rahbomyolosis (internal heating and buildup of acid in the heart muscle). This condition caused several major organs to fail resulting in death.

[4] "Exposure/Contact with" follows NFIRS 4.0 definitions under "Cause of Fatal Injury".

Table 2 Cause of Fatal Injury

	Number	Percent
Stress or Overexertion	47	50.0%
Struck by or Contact with Object	33	35.1%
Caught or Trapped	7	7.4%
Exposure	5	5.3%
Fell or Jumped	2	2.1%
TOTAL	94	100%

died from an asthma attack after being exposed to toxic substances at a fire, and another died after inhaling too much smoke while attempting a rescue at a structure fire. Two other firefighters died from exposure. One died as a result of burns after he was caught in a flashover, and the other died from cardiac arrest/pulmonary edema after being exposed to a cloud of unknown chemical vapors at a commercial structure fire.

Two firefighters died as a result of falls. One firefighter slipped and fell down a fire pole about 20 feet. Another firefighter died after falling and striking his head.

NATURE OF FATAL INJURY

Table 3 and Figure 6 show the distribution of the 94 deaths by the medical nature of the fatal injury or illness. The leading nature of death in 1996 was heart attacks, which accounted for 46 firefighter fatalities. Two of the heart attacks occurred while exercising, and one occurred during an agility/stress test. There were 24 firefighters who suffered heart attacks at fire scenes[5] and five who suffered heart attacks enroute to or returning from calls. Three heart attacks occurred at EMS or rescue incidents. Three others occurred during fund-raisers, two while asleep at the fire station, one while repairing fire apparatus, one just following a fire, and four during training.

Internal trauma was the second leading nature of death, responsible for 32 deaths (up nine from 1995). This total includes 18 firefighters who were involved in vehicle accidents, four who were hit by vehicles while on the emergency scene, and six firefighters who were victims of violence. Four other firefighters died as a result of internal trauma—one firefighter died after falling down a fire pole hole, one died when fireworks detonated at a Fourth of July celebration, one was crushed by a collapsing wall, and one was killed when a tree fell on the fire apparatus.

[5] One firefighter included in this total died of cardiac arrest as a result of an exposure at a hazmat incident.

Table 3 Nature of Fatal Injury

	Number	Percent
Heart Attacks	46	48.9%
Internal Trauma	31	33.0%
Asphyxiation (includes drowning)	10	10.6%
Electrocution	3	3.2%
Dehydration	1	1.1%
Stroke/Seizure	1	1.1%
Burns	1	1.1%
Other (sickle cell anemia)	1	1.1%
TOTAL	94	100%

Figure 6 Nature of Fatal Injury

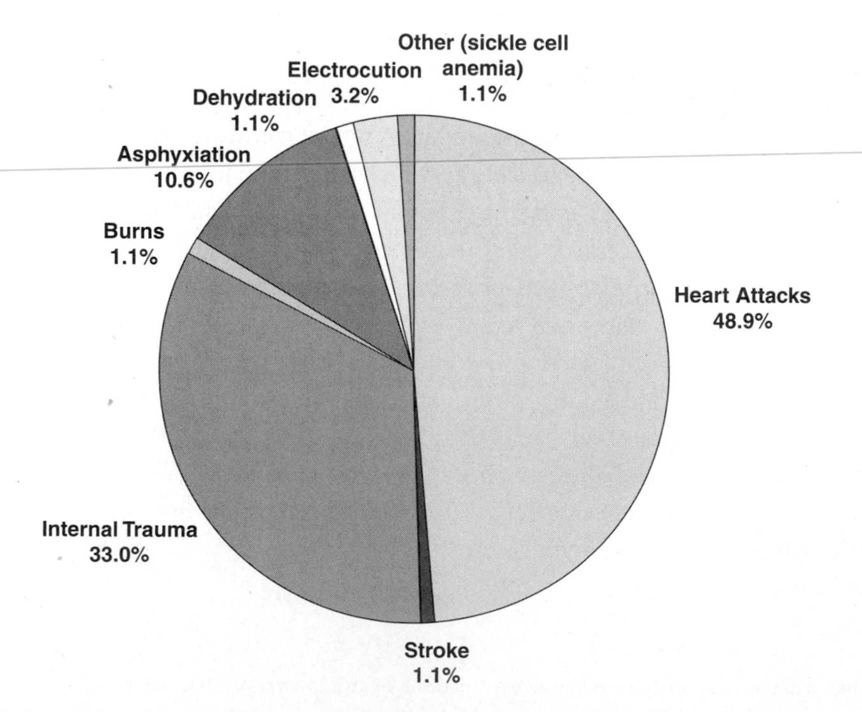

Asphyxiation was the third leading medical reason for firefighter deaths, responsible for 10 deaths (50% less than 1995). Of the ten firefighter deaths, eight resulted from carbon monoxide poisoning or inhalation of smoke or superheated gases during structural firefighting. All of these eight deaths occurred when the firefighters became caught and trapped by rapidly spreading fires or structural collapses. Two other firefighters died as a result of asphyxiation—one when he took his SCBA mask off in an oxygen deficient atmosphere and another when he became trapped and drowned after wrecking his car in a lake.

Three firefighters died from electrocution—one whose SCBA became caught in downed power lines, one who came into contact with downed power lines at an automobile accident, and one who had power lines fall on him when a power pole broke and fell to the ground.

Only one of the 94 firefighter fatalities was attributed to burns. The firefighter was caught in a flashover in a two-alarm fire in an apartment.

The medical causes of death for the final three were dehydration, stroke, and organ failure due to a preexisting condition of sickle cell anemia.

FIREFIGHTERS' AGES

Figure 7 shows the distribution of firefighter deaths by age and cause of death. Younger firefighters were more likely to have died as a result of traumatic injuries from an apparatus accident or after becoming caught or

Figure 7 Age & Cause

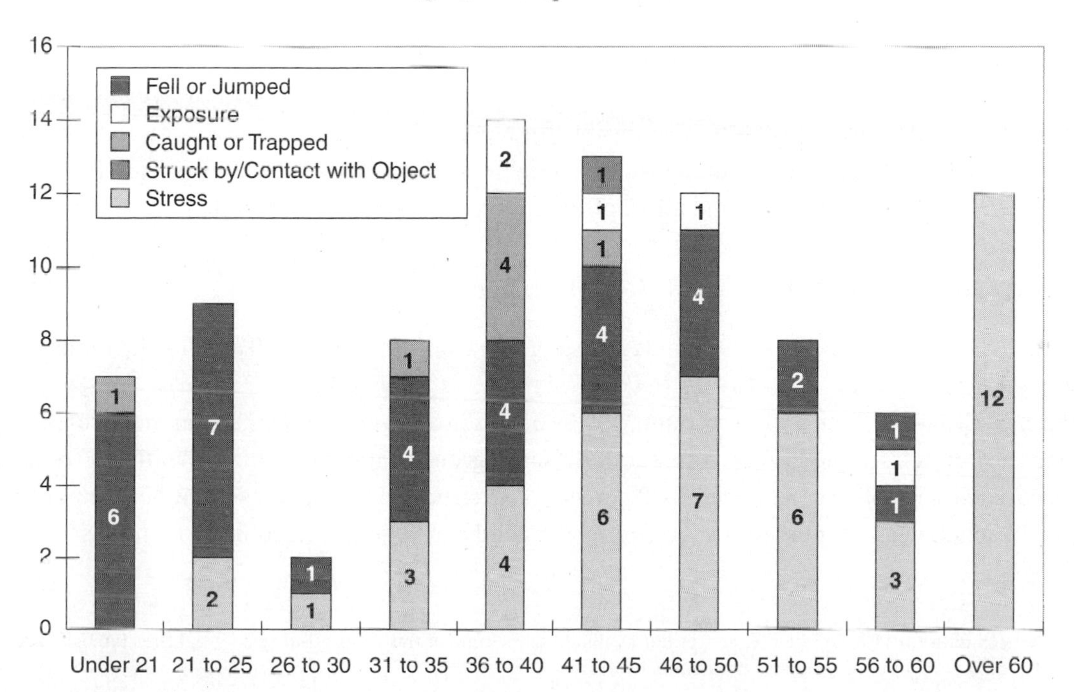

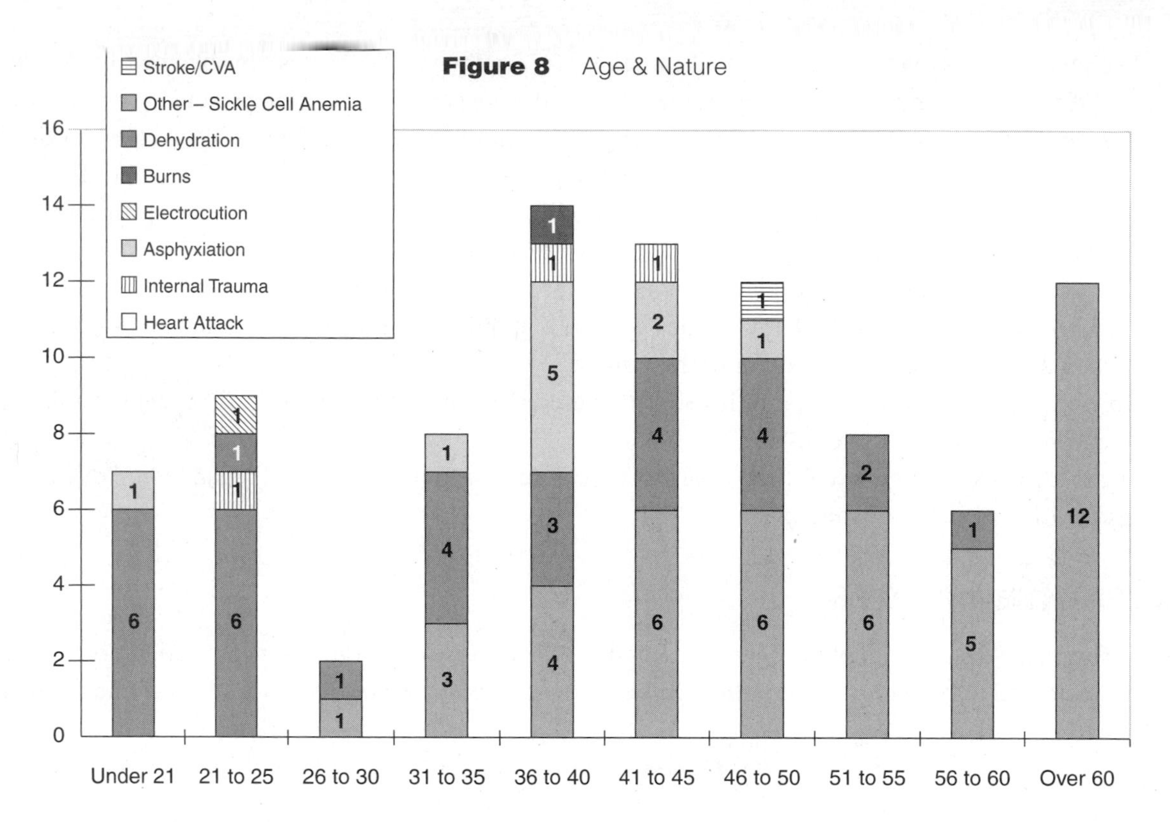

trapped during firefighting operations. Stress was shown to play an increasing role in firefighter deaths as age increased. This is also reflected in Figure 8, which shows the distribution of deaths by age and medical nature of injury. Trauma and asphyxiation were responsible for most of the deaths of younger firefighters, while heart attacks were much more prevalent among older firefighters. Heart attacks accounted for 16 of the 26 firefighters who were over 50 years old, and all 12 of the firefighters over 60 years old.

FIREGROUND DEATHS

There were 38 fireground deaths in 1996, a decrease of two from 1995. Figure 9 and Table 4 show the distribution by fixed property use.

Property Type—31 of the 38 fireground deaths occurred at structure fires. As in most years, residential occupancies accounted for the highest number of these fireground fatalities, with 19 deaths (50 percent). Residential occupancies usually account for 70–80 percent of all structure fires and a similar percentage of the civilian fire deaths each year, 50 percent of the firefighter deaths in 1996 occurred in residential structures.[6] The

[6] Complete NFIRS data for 1996 fire incidence was not available at the time of this report, but residential fires typically account for between 70 and 80 percent of all civilian fatalities each year.

Figure 9 Fixed Property Type

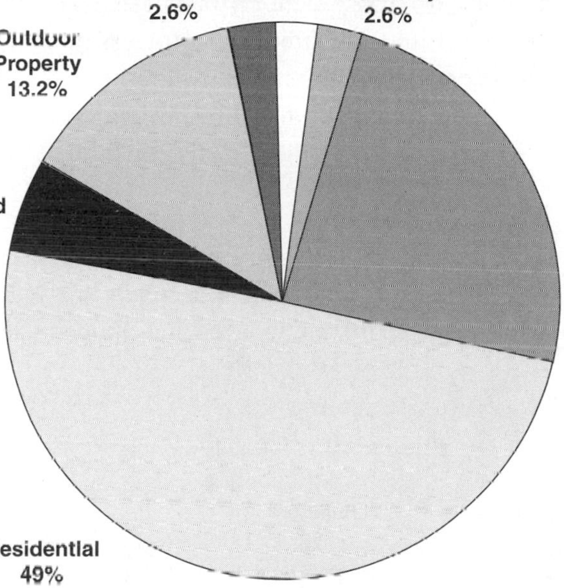

Table 4 Fixed Property Use for Fireground Deaths

	Number	Percent
Residential	19	50.0%
Commercial	9	23.7%
Outdoor Property	5	13.2%
Street/Road	2	5.3%
Storage	1	2.6%
Public Assembly	1	2.6%
Manufacturing	1	2.6%
TOTAL	38	100%

frequency of firefighter deaths in relation to the number of fires is much higher for non-residential structures. One firefighter died in 1996 in a storage occupancy compared to six in 1995. Nine firefighters died in commercial structure fires, one in public assembly, and one in manufacturing.

Outdoor properties and "street/road" accounted for a total of seven deaths. Four out of the five outside property deaths were heart attacks. The fifth was a firefighter who was pinned between two trucks at a wildland fire. The two street/road deaths consisted of a firefighter who was electrocuted when a power line came down at a pole fire, and one who had a heart attack while directing traffic at an emergency scene.

TYPE OF ACTIVITY

Figure 10 and Table 5 show the activities the 38 firefighters were engaged in at the time they sustained their fatal fireground injuries or illnesses. There was a substantial decrease this year compared to last year in the number of firefighters who died while engaged in traditional engine company duties of fire attack and advanc-

Figure 10 Type of Activity

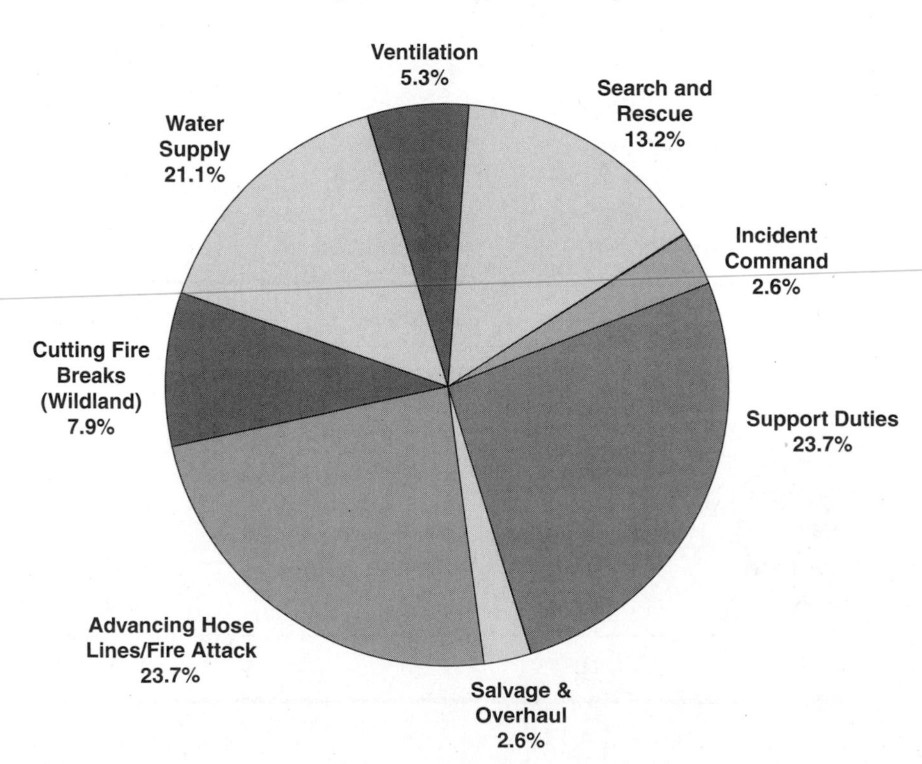

Table 5 Type of Activity for Fireground Deaths

	Number	Percent
Advancing Hose Lines/Fire Attack	9	23.7%
Support Duties	9	23.7%
Water Supply	8	21.1%
Search and Rescue	5	13.2%
Cutting Fire Breaks (Wildland)	3	7.9%
Ventilation	2	5.3%
Incident Command	1	2.6%
Salvage & Overhaul	1	2.6%
TOTAL	38	100%

ing hose lines (decrease of 14). Nine firefighters died while performing these fireground operations, including six who died from asphyxiation after becoming trapped by rapid fire spread or structural collapse while advancing hose lines. Three other firefighters suffered heart attacks while performing similar functions. Eight firefighters died while performing water supply operations on the fireground—five from heart attacks, one from electrocution, one from being pinned between two fire apparatus, and one from being struck by a passing motorist.

Traditional truck and ladder company duties accounted for eight deaths. Search and rescue operations in burning structures were being conducted when five of these deaths occurred, no change from 1995. Three of the search and rescue deaths were from heart attacks, one from burns after being caught in a flashover, and one from asphyxiation. Two firefighters died while ventilating structure fires, one from a heart attack and one from internal trauma when a wall collapsed. One firefighter was electrocuted during salvage and overhaul operations when a power line came into contact with his SCBA.

Nine firefighters died while performing support functions or standing by on the fireground—seven from heart attacks, one from asphyxiation, and one from internal trauma.

Cutting fire lines to contain grass, brush, and forest fires accounted for three firefighter fatalities. All three died as a result of heart attacks.

One incident commander suffered a fatal cardiac arrest/pulmonary edema at fire incidents.

TIME OF ALARM

The distribution of 1996 deaths according to the time of day when the incidents were reported is shown in Figure 11 (49 times were not reported). The highest number of fireground deaths occurred for alarms that were received between 2300 and 0059. The second highest number was a tie between 1700–1859 and 1100–1259. There were no fireground deaths between the hours of 0700–0859.

Figure 11 Time of Death

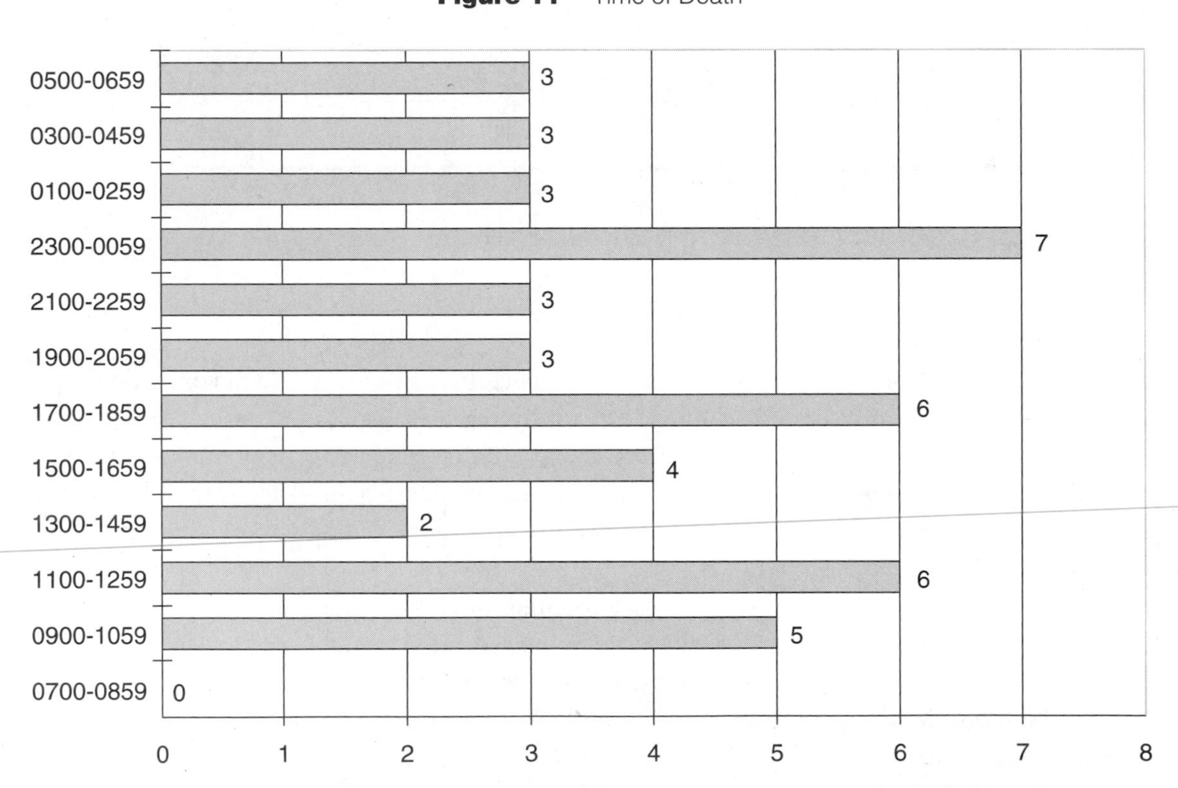

MONTH OF THE YEAR

Figure 12 illustrates firefighter fatalities by month of the year. Firefighter fatalities peaked in January and April. Other high months were recorded in August and October. The early summer months (May, June, and July) were among the lowest months. (Conversely the number of residential fires peaked during the winter and was lowest during June and July.)

Figure 12 Deaths by Month of the Year

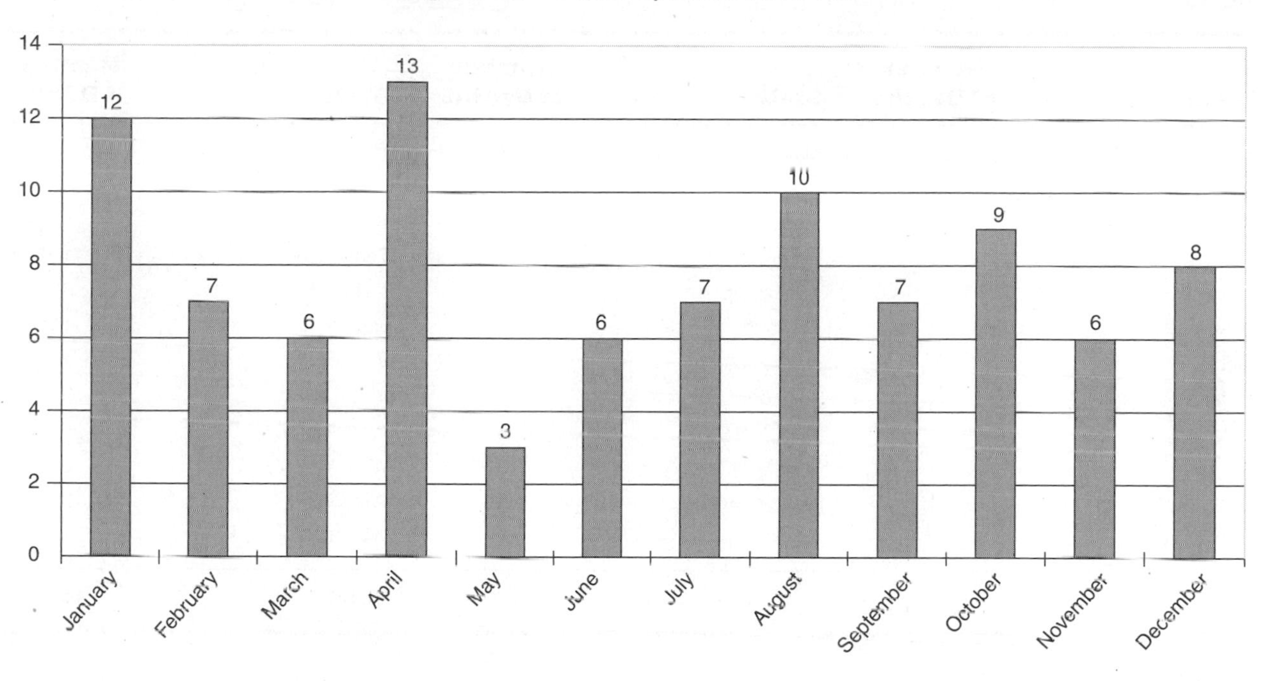

STATE AND REGION

The distribution of firefighter deaths by state is shown in Table 6.[7] Thirty-four states had at least one firefighter fatality. New York led with 13 deaths. Figure 13 shows the firefighter fatalities divided by region of the country and whether they were career structural, volunteer structural, or career or seasonal wildland firefighters.

ANALYSIS OF URBAN/RURAL/SUBURBAN PATTERNS IN FIREFIGHTER FATALITIES

The US Bureau of the Census defines "urban" as a place having a population of at least 2,500 or lying within a designated urban area. Rural is defined as any community that is not urban. Suburban is not a census term but may be taken to refer to any place, urban or rural, that lies within a metropolitan area defined by the Census Bureau, but not within one of the central cities of that metropolitan area.

[7] This list attributes the deaths according to the state where the fire department or unit is based, as opposed to the state where the death occurred. They are listed by those states for statistical purposes, and for the National Fallen Firefighter Memorial at the National Fire Academy.

Table 6 1996 State with On-Duty Firefighter Fatalities

State	Number of Deaths	State	Number of Deaths	State	Number of Deaths
Alabama	2	Maine	1	Oklahoma	4
Arkansas	1	Maryland	5	Pennsylvania	6
Arizona	1	Massachusetts	2	South Carolina	4
Connecticut	2	Michigan	2	Tennessee	1
Georgia	4	Mississippi	4	Texas	5
Hawaii	2	Missouri	1	Utah	1
Illinois	4	Nebraska	2	Vermont	1
Indiana	4	Nevada	1	Virginia	3
Iowa	1	New Jersey	5	Washington	1
Kansas	1	New York	9	West Virginia	2
Kentucky	1	North Carolina	3	Wisconsin	2
Louisiana	2	Ohio	3	Wyoming	1
				Total:	**94**

Figure 13 Firefighter Deaths by Region 1996

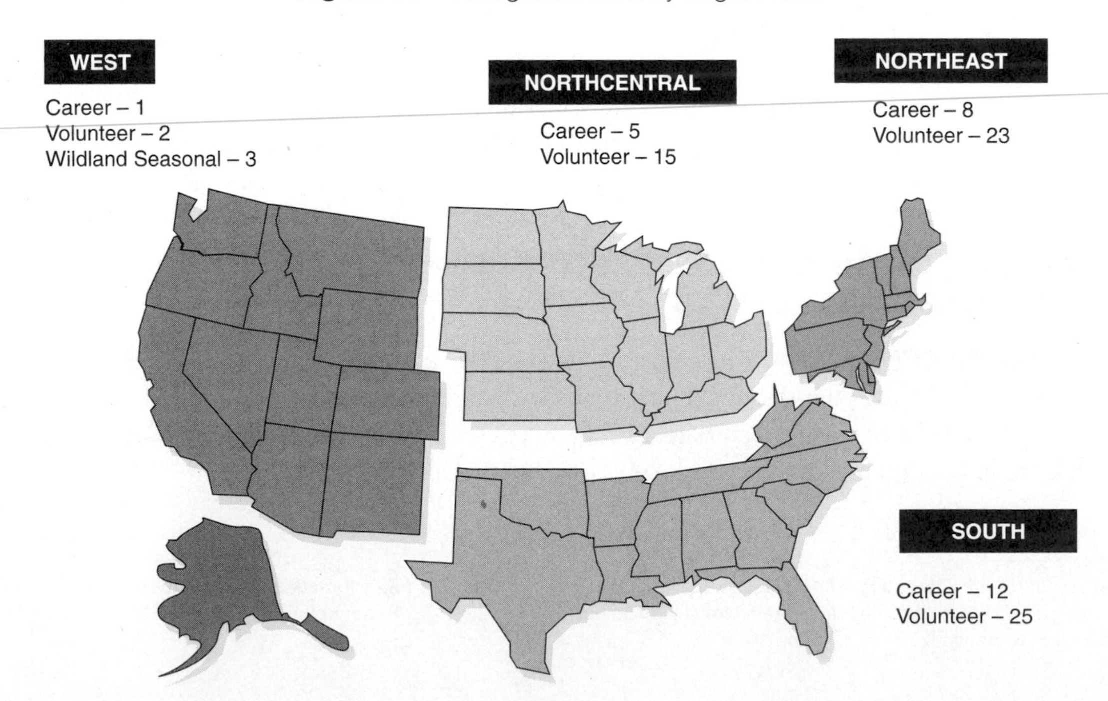

Fire department areas of responsibility do not always conform to the boundaries used for the census. For example, fire departments organized by counties or special fire protection districts may have both urban and rural coverage areas. In such cases, it may not be possible to characterize the entire coverage area of the fire department as rural or urban, and firefighter deaths were listed as urban or rural based on the particular community or location in which the fatality occurred.

The following patterns were found for 1996 firefighter fatalities. These are estimates based upon population and area served reported by the fire departments.

Table 7

	Urban/Suburban	Rural	Federal or State Parks/Wildland	Total
Firefighter Deaths	61	30	3	94

SPECIAL TOPICS

HOMICIDES AND VIOLENCE IN THE WORKPLACE

Violence towards emergency services providers is a growing concern. In 1996, six firefighters were murdered while on duty, including four in a single shooting incident. In numerous other cases, firefighters have been the victims of violence perpetrated by random attackers, by the citizens they have attempted to serve, and by fellow workers. The fire service is no longer immune to these events and trends which have been affecting other segments of society for many years.

National statistics on workplace violence are a cause for increasing alarm. Homicide has become the third leading cause of death in the workplace. Murders in the workplace are also the fastest growing type of murder in the United States. The murder rate of supervisors has doubled since 1985. Statistical information shows that the murder rate for public sector employees is more than twice that of their private sector counterparts.[8]

The fire service does not adequately track violence directed towards employees and volunteers. Reports of violence vary widely by state and locality. No NFPA standard deals directly with the issue of reducing violence towards emergency workers, except the portion of NFPA 1500, Standard on Fire Department Occupational Safety and Health Program, that references response to civil disturbances. The Occupational Safety and Health Administration (OSHA) has a policy that employers must provide a workplace free of violence, however applying this policy to firefighters is difficult, because they, like paramedics and police officers, work on the streets in good and bad neighborhoods, under stressful situations, and without the safety that can be engineered into an individual worksite.

The six murdered firefighters represent 6.4 percent of firefighter deaths in 1996. This contrasts with the traumatic (includes asphyxiation) deaths of only seven firefighters while engaged in interior firefighting during the same year.

Six Murders—In the most highly publicized workplace violence incident in 1996, six Jackson (MS) Fire Department career fire officers were shot by a fellow firefighter while attending a staff meeting on April 24. Four of the officers were wounded fatally. The firefighter murdered his estranged wife earlier in the day, and went to the city's central fire station, where he barged into the staff meeting and started shooting. He also tried to force his way into the Fire Chief's office. He was apprehended by police after a chase and shoot-out.

Fellow firefighters indicated that the individual had previously expressed his hatred of chief officers. Widespread racial tensions in the department are also thought to have contributed to the motivation for the

[8] Barrett, Stephen. "Protecting Against Workplace Violence," *Public Management,* August 1997. Pages 9–12.

shooting incident. Police later linked one of the weapons used by the firefighter to two previously unsolved murders in Jackson.

In an incident that occurred on January 7, 1996 in Pleasantville, New York, a firefighter was murdered while enroute to his station. The volunteer firefighter was walking to the station during a blizzard to report for storm duty, but he never arrived. His body was discovered several hours later. His throat had been slashed and he had been stabbed and bludgeoned to death. The police have not made any arrests in this case.

In the third incident, an 18 year old volunteer firefighter was shot to death in Indiana when he was the first to arrive at the scene of a motor vehicle accident involving a car and a motorcycle. The driver of the car shot the motorcyclist and the firefighter, killing them both.

Preventative Strategies—OSHA separates workplace violence into three categories based upon the perpetrators' links to the workplace. Type I violence occurs when an employee is the victim of a random act of violence from an offender with no apparent ties to the workplace. This could have been the case in the murder of the firefighter in New York. Type II violence occurs when the attacker is a recipient of the victim's services, such as a patient attacking an EMS crew. This was the case in Indiana. Type III violence occurs when the attacker has a direct relation to the workplace, such as a fellow employee, former employee, or relative or acquaintance of an employee. This was the case in Jackson, Mississippi.

Firefighters are at increased risk for all three types of violence. Stations, with few exceptions, are open buildings widely visited by the public. The nature of firefighting, rescue, and EMS work brings the employee or volunteer into very dangerous areas of their communities, often directly into the path of domestic feuds or other violent situations. Customers may be mentally ill or under the influence of drugs or alcohol, making their actions erratic, unpredictable, and violent. Compounding these external dangers are the pressures felt by tired and stressed co-workers due to work or personal problems. Risks are increased when families or acquaintances become involved.

There are strategies that departments can take to reduce violence or death to workers. Random violence may be reduced through improvements in station design and security. Firefighters should be made aware that they can be victims while on-duty, and that they may be targeted because of their job and profile in the community, or because they work at all hours in all types of areas. Violence towards workers by patients and bystanders at incident scenes may be reduced by training firefighters in verbal de-escalation methods, self-defense tactics, and survival training, similar to the training that police officers receive. They should be trained to recognize and avoid a dangerous situation or conflict in the first place, to try to verbally defuse it if they become involved, and to defend themselves to survive or escape an attack. Additionally, police should escort units to high crime areas and certain types of incidents. Body armor should be made available to any firefighters who respond to high crime areas. Firefighters and EMS workers should be clearly identified as different from police officers. Departments should consider distinct uniforms, ones that clearly identify personnel as firefighters when they are not in their turnout gear, but do not look like local police uniforms.

In the case of type III violence by coworkers, departments across the United States should consider several improvements in their recruitment and evaluation processes for employees or volunteers. Few departments do psychological profiling of members in addition to thorough background checks. This should be expanded, to help identify applicants who may not be suited for the particular stresses of a firefighting career. Employees should be monitored throughout their careers for emotional and mental fitness and stability. Stress reduction and wellness programs should be expanded to help prevent firefighters from succumbing to mental stressors. Employee assistance programs and mental health professionals should be available on a confidential basis for employees who need psychological assistance. Conflict management training and sensitivity training should be provided at all levels, not just for managers. Confidential reporting systems should be developed to help overcome the culture of non-reporting of fellow workers who become a threat to themselves or to others. Mechanisms for psychologically assisting discharged employees through their transition into other jobs should be considered.

Firefighting is dangerous enough without the additional stress of threatened violence. By focusing on training firefighters to operate more safely in dangerous situations, by properly screening prospective employees, and by providing continual psychological counseling for firefighters and their families, the fire service can hopefully prevent some of these tragic deaths in the future.

FIREFIGHTER HEALTH AND WELLNESS

1996 saw a continued positive emphasis on firefighter health and wellness throughout the fire service. A highlight in this area was the cooperative effort to develop firefighter physical fitness and wellness programs agreed to by the International Association of Fire Fighters (IAFF) and the International Association of Fire Chiefs (IAFC). In concert with 10 fire departments throughout the United States and Canada, the IAFF and IAFC will jointly design a program that can be used by fire departments everywhere to improve the health and wellness of firefighters.

The development and implementation of health and wellness programs for firefighters is instrumental for reducing the annual number of firefighter injuries and fatalities. Under ideal conditions, firefighting is a strenuous task requiring an above-average level of physical fitness. Additional demands are placed on firefighters by the many inherent stresses of the job, which can affect their physical and emotional health and well being. The nature of firefighters' unusual work and eating schedules also requires that attention be paid to wellness efforts like dietary education and modification, body composition testing, and stress reduction.

Data collected for 1996 indicates that improvements in firefighter health and wellness are still sorely needed: 46 fatalities occurred in 1996 as a result of heart attacks, of which 45 were attributed to stress or overexertion, with one attributable to cardiac arrest subsequent to a hazardous materials exposure. Heart attacks were the leading cause of firefighter deaths, representing 48.9% of all 1996 fatalities.

Heart attacks struck career (13), volunteer (31), and wildland seasonal (2) firefighters with ages ranging from 29 to 78. These firefighters were performing a variety of activities including: advancing hose lines, search and rescue, ventilation, fireground support, training, water supply, cutting fire breaks, and station duties, among others.

The 1996 data clearly suggests that there is still a great deal of work to be done in the firefighter health and wellness arena. Both career and volunteer firefighters stand to benefit from the cooperative effort currently being undertaken by the IAFF and IAFC. Fire departments will also benefit from the reduced casualty rates that can be expected with improved firefighter health and wellness.

VEHICLE ACCIDENTS

In recent years, a great deal of emphasis has been placed on the safe operation of fire department vehicles while responding to, or returning from, incidents. Emergency vehicle operator courses (EVOC) and defensive driving classes are increasingly being required for drivers of fire department vehicles. Commercial driver's licenses (CDLs) or special vehicle operator's endorsements may also be required by individual states, although many states still exempt drivers of fire department apparatus from these requirements. Some states offer comprehensive driver/operator courses in a modular format that allows them to be presented to both career and volunteer personnel, although the completion of such a course may or may not be required for vehicle operators. Several fire departments have implemented response procedures designed to minimize the amount of emergency or "lights and siren" driving required, by prioritizing response levels such that later-arriving units respond in a routine or "non-lights and siren" mode.

Despite the recent emphasis on vehicle safety throughout the fire service, the 1996 data indicates that more work needs to be done in this area, as 22 firefighter deaths occurred as a result of vehicle accidents. Although the number of deaths from vehicle accidents is less than last year (down 8 deaths from 1995) they still represent 22.4% of all 1996 firefighter fatalities. For 1996, as in every year since 1993, vehicle accidents remain the second leading cause of firefighter fatalities.

The causes of the enroute fatalities were varied[9]; eight firefighters were killed as a result of trauma incurred during fire apparatus accidents, five of those involved apparatus rollovers. Eight firefighters were killed while operating their personal vehicles enroute to emergency calls, including one who drowned in a lake after his vehicle wrecked.

All 16 of the firefighters killed as a result of vehicle accidents were volunteers.[10] This data may indicate that further efforts are needed to train and qualify volunteer firefighters as emergency vehicle operators. The nature of the volunteer service may make it difficult for vehicle operators to maintain proficiency or familiarity with the unique driving requirements of fire apparatus. Defensive driving and vehicle operator training is also important for volunteers who respond to emergencies in their personal vehicles. Such training may have the added benefit of reduced insurance premiums for members and the fire department.

In order to reduce the number of firefighter fatalities related to vehicle accidents, firefighters must remember that emergencies will only worsen unless the fire department response is conducted with a high

[9] Five firefighters that suffered from fatal heart attacks while responding to or from incidents are not mentioned in this section.

[10] Five firefighters that suffered from fatal heart attacks while responding to or from incidents are not mentioned in this section.

degree of safety, ensuring that personnel and equipment arrive unharmed and ready to do the job. The development of safe firefighter attitudes coupled with improved training and certification requirements should help reduce the number of vehicle accidents that occur while responding to, or returning from, incidents. Reducing the number of fire department vehicle accidents should contribute to a reduction in the number of firefighter casualties resulting from these accidents.

CONCLUSIONS

The analysis of firefighter deaths in 1996 indicates that the overall long-term trend toward fewer firefighter fatalities is continuing. The 94 fatalities in 1996 are the third lowest recorded, and only the third time the total number of fatalities has dropped below 100, all within the last four years.

Stress-induced heart attacks continue to remain the number one cause of firefighter deaths.[11] Health and fitness programs should help to reduce these numbers in the long term. Better screening for high risk firefighters through medical exams may help to prevent some deaths in the short term, by identifying firefighters who are unfit for strenuous duty or who may be at high risk for heart disease. Many factors that place firefighters at high risk for heart attacks are controllable, such as better nutrition, not smoking, and exercise.

Figure 14 Stress vs. Actions

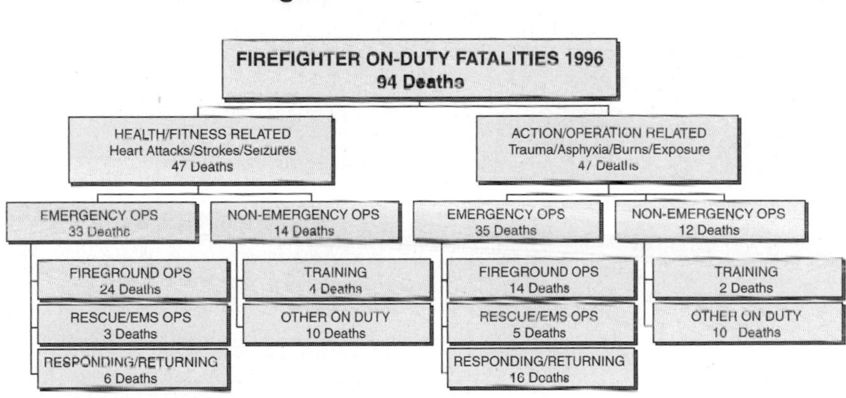

Six firefighters died of asphyxiation in structure fires, all the result of a structural collapse. These and several other incidents reinforce the need for proper size-up, continual progress reports, and a working accountability system at all incidents to keep track of personnel. An accountability system should track all members, including who and where they are, who they are working for, what they are doing, and how long they have been doing it.

[11] Figure 14 shows the relation between stress-related and action-related firefighter deaths.

Perhaps the best way to reduce fireground fatalities is through the adoption of more stringent building codes, better fire prevention, and strict code enforcement.

The institution of rapid intervention teams into emergency operations may help save more firefighters' lives. PASS devices should be used at all fire incidents, to improve the chances of being alerted about, and locating, a downed firefighter.

Responding to incidents continues to claim too many lives. In 1994, personal vehicle accidents as a cause of fatalities were almost eliminated, however, in 1995 the number of vehicle accident deaths rose and continued into 1996. All fire departments should have a policy regarding driver training for responding to emergency incidents, and all drivers should use caution when approaching intersections. Also, seat belts can only save lives if they are worn; they were not used in several 1996 fatal accidents.

As in last year's analysis, the 1996 statistics indicate that risk management, in the form of assessing firefighters' health risks, sizing up fireground conditions, and evaluating hazards at special rescue scenes, is a key to reducing on-duty firefighter fatalities even further.

Firefighting is a dangerous occupation and the additional stress of threatened violence is not needed. Fire departments, like many other industries are doing, should develop training to teach firefighters to operate more safely in dangerous situations, properly screen prospective employees, and provide continual psychological counseling for firefighters and their families. As a result, the fire service can hopefully prevent some of these tragic deaths in the future.

1996 INCIDENT SUMMARY

INCIDENT 1

FDNY, New York, NY
On January 5th, James B. Williams, a career Firefighter with FDNY, died from burns sustained during a 2-alarm fire rescue at an apartment building in Queens, New York. Unaware that the occupants of the apartment had already left, he and four other firefighters were searching for victims and fighting the fire when they were engulfed in flames after breaking through a door.

INCIDENT 2

West End Fire Company, Stowe, Pennsylvania
On January 5th, William R. Favinger, Sr., a volunteer for the West End Fire Company, suffered a fatal heart attack while returning from an automatic alarm. He collapsed in the station while filling out the roster. CPR was immediately started and paramedics were called.

INCIDENT 3

Owego Fire Department, Newark, New York
On January 6th, volunteer Firefighter Guy R. Pollard suffered a fatal heart attack while performing pump operations at a suspected house fire on a mutual call with the Owego Fire Department. After determining that the house's chimney was stuffed, trucks begin preparing to leave when Firefighter Pollard suffered the heart attack. He was transported to Wilson Memorial Regional Medical Center where he was pronounced dead on arrival.

INCIDENT 4

Pleasantville Fire Department, Pleasantville, New York
On January 7th, Firefighter Thomas Dorr was killed as a result of multiple injuries (stab wounds) while responding to the station on foot during a snow storm. He was walking to the firehouse for storm duty and was attacked at some point.

INCIDENT 5

Rockaway Borough Fire Department, Rockaway, New Jersey
On January 7th, volunteer Firefighter Willard Hopler suffered a massive MI while operating an aerial apparatus at the scene of a chimney fire. Despite resuscitative efforts that were performed by his crew, he was pronounced dead at Northwest Covenant Medical Center.

INCIDENT 6

Pecatonica Fire Protection District, Pecatonica, Illinois
On January 13th, Chief Dale Zimmerman died in an attempt to rescue two men who were overcome by fumes in a grain bin. During the rescue, alarm bells went off warning that another firefighter's air tank was getting low. The Chief went over to change the tank, but his mask was fogging up. He took off his mask in order to change the tank. The other fireman then left, but saw the chief having problems in his attempt to rescue the two men. The two men were rescued, but the chief eventually died from asphyxia from the carbon monoxide poisoning.

INCIDENT 7

Cairo Fire Department, Cairo, Georgia
On January 18th, Firefighter Marcel Glenn died while fighting a structure fire. Firefighter Glenn was ventilating a house fire by breaking the windows with a fire hose. After he had ventilated two windows, he turned around and collapsed. EMS was called and he was taken to the hospital where he was pronounced dead due to cardiac arrest.

INCIDENT 8

Citizens Hose Company #1, South Renova, Pennsylvania
On January 19th, volunteer Firefighter Reed Morton Sr., suffered a heart attack while directing traffic and assisting evacuees at a fire (fire policeman).

INCIDENT 9

Wildwood Fire Association, Alger, Michigan
On January 19th, volunteer Firefighter Robert Haggadone was struck by a passing motorist while working a hydrant at a house fire. He died after being in a coma for 7 1/2 months.

INCIDENT 10

Dallas Fire Department, Dallas, Texas
On January 22nd, recruit Firefighter Jerald Dibbles died during his second day at the training academy. He had a pre-existing condition of sickle cell anemia (trait). Because of this condition he went into a state of rahbomyolosis (internal heating and buildup of acid in the heart muscle). This condition caused several major organs to fail resulting in death.

INCIDENT 11

Livingston Fire Department, Livingston, Texas
On January 26th, volunteer Firefighter Dale Burkhalter died in a car accident while returning from a fire incident. The accident occurred at four in the morning at a dark and unlighted intersection. The road conditions were wet and there were patches of fog. In an attempt to cross a major highway, the firefighter's car was struck on the driver's side.

INCIDENT 12

Lake Murray Village Fire Department, Ardmore, Oklahoma
On January 31st, volunteer Marvin Maphes drove a tanker to the fire scene and suffered a fatal heart attack on arrival.

INCIDENT 13

Kauai County Fire Department, Kauai, Hawaii
On February 1st, Firefighter Steven Gushiken got up in the morning (4:30 a.m.) while still on-duty and went for a walk in the park adjacent to the station (normal morning routine). When the rest of the shift woke up later they found him unconscious on the ground. Coworkers unsuccessfully attempted to revive him.

INCIDENT 14

FDNY, Brooklyn, New York
On February 5th, Firefighter Louis Valentino became trapped when the roof of an auto body shop in East Flatbush-Brooklyn collapsed. Fifteen other firefighters were injured at this 3-alarm blaze. Firefighter Valentino died less than an hour after the fire started at about 3:40 p.m.

INCIDENT 15

Clarksville Fire Department, Clarksville, Virginia
On February 5th, Firefighter Corey Morgan died in a motor vehicle accident while responding to a fire call.

INCIDENT 16

Ridgefield Boro Fire Department, Ridgefield, New Jersey
On February 11th, Firefighter Michael McLaughlin died of an apparent heart attack after arriving on the scene of a small fire in a laundromat. Firefighter McLaughlin experienced head trauma when he fell on the scene and knocked his head against the fire engine. This trauma in turn resulted in cardiac arrest.

INCIDENT 17

Enville Fire Department, Marietta, Oklahoma
On February 11th, volunteer Firefighter Raymond Vinson fought a grass fire for approximately seven hours in the morning, when he was called out again for another grass fire. This incident lasted about five hours. He died of a heart attack after returning from the incident.

INCIDENT 18

I.X.L. Fire Department, Castle, Oklahoma
On February 23rd, Firefighter Nathaniel Quinn went into cardiac arrest while fighting a wildland fire near Okemah. The wildfires blackened up to 30,000 acres, and destroyed 43 homes in 10 Oklahoma counties.

INCIDENT 19

Ash Township Volunteer Fire Department, Carleton, Michigan
On February 24th, Firefighter Francis Ploeger arrived at the scene of a two-alarm barn fire. While pulling a hose from the fire truck, he collapsed due to a heart attack. CPR was initiated at the scene, and he was taken to a hospital where he was pronounced dead later that evening.

INCIDENT 20

Allentown Road Volunteer Fire, Fort Washington, Maryland
On March 2nd, volunteer Firefighter Leonardo Maguidad went into cardiac arrest at the station while on duty.

INCIDENT 21

Tomah Volunteer Fire Department, Tomah, Wisconsin
On March 7th, volunteer Firefighter Dennis McGary suffered a fatal heart attack after returning from a house fire. After returning, Firefighter McGary was putting away equipment and preparing firehouse items when the heart attack occurred.

INCIDENT 22

Alexandria Volunteer Fire Department, Alexandria, Nebraska
On March 8th, Firefighter Vinton Durflinger collapsed and died due to a heart attack after checking out a suspected house fire.

INCIDENT 23

Golden City Volunteer Fire Department, Golden City, Missouri
On March 13th, Firefighter Norman Manka was operating the pump at a grass fire when he collapsed and died due to a heart attack.

INCIDENT 24

Chesapeake Fire Department, Chesapeake, Virginia
On March 18th, Firefighters Frank Young and John Hudgins died while battling a blaze in an Advanced Auto Parts Store. Both firefighters became trapped by fire when the truss roof collapsed on top of them. The firefighters were found in the rear of the structure some time later.

INCIDENT 25

Granville Fire Protection District, Granville, Illinois
On April 7th, Firefighter Robert Duvall suffered a fatal heart attack while fighting a house fire in Hennepin, IL.

INCIDENT 26

Almena V. Fire Department, Almena, Kansas
On April 8th, Firefighter Norman Adams died from an asthma attack after engaging in support duties for 9 1/2 hours at an industrial fire in a plant that makes aluminum products.

INCIDENT 27

Grant County Fire District 5, Moses Lake, Washington
On April 8th, Firefighters Boster and Fowler were responding to a reported mobile home fire when they rounded a corner too quickly and the tanker they were in rolled onto its side. Firefighter Boster was killed and Fowler was treated for broken ribs and other minor injuries. It was not reported whether the firefighters were wearing seatbelts.

INCIDENT 28

Harlan Township Fire & Rescue, Pleasant Plain, Ohio
On April 10th, Firefighter Terry Leasher died of internal injuries due to a motor vehicle accident. He was on his way to the station to perform truck inspections.

INCIDENT 29

Schenectady City Fire Department, Schenectady, New York
On April 19th, Firefighter Donald Collins was hooking up a hose at the scene of a vacant house fire around 2 a.m. when he went into cardiac arrest, collapsed, and was taken to the hospital and pronounced dead.

INCIDENT 30

Wayne Fire Department, Wayne, Oklahoma
On April 19th, Firefighter Mathew Hatcher died of abdominal injuries after being pinned between two fire trucks at a grass fire. He was at the rear of one truck starting the pump when a second truck pinned him.

INCIDENT 31

Omaha Fire Department, Omaha, Nebraska
On April 23rd, Firefighter Goessling was killed when the roof collapsed on him at a 4-alarm fire in a commercial building (Dollar General). A 15-yr old has been arrested on suspected arson.

INCIDENT 32

Division of Forestry & Wildlife, Wailuko, Hawaii
On April 23rd, Firefighter Mark Clark died while participating in a chainsaw (tree felling) training class. He was clearing brush around a tree when he put his saw down, collapsed, and died of a heart attack.

INCIDENT 33

Jackson Fire Department, Jackson, Mississippi
On April 24th, Captain Stanley Adams, Captain Don Moree, District Chief Willie Craft, and District Chief Rick Robbins were shot to death during a meeting among district chiefs. They were killed by a disgruntled firefighter who went on a rampage killing five people including his wife. Two other people were also injured in this incident.

INCIDENT 34

Antioch Volunteer Fire Department, Beebe, Arkansas
On April 26th, Firefighter Robert Pemberton was killed in an apparatus accident while enroute to a reported structure fire. He was ejected from the driver's seat after the truck failed to negotiate a turn and then overturned several times. He was pronounced dead at the scene.

INCIDENT 35

Atlanta Fire Department, Atlanta, Georgia
On May 1st, Firefighter Robert Hamler suffered a stroke at the fire station. He had been out inspecting fire hydrants when he started to feel poorly. He was then taken to the hospital.

INCIDENT 36

Mahwah Township Fire Department, Mahwah, New Jersey
On May 25th, Assistant Chief Kevin Malone died of a heart attack at home after returning from a false alarm. He had complained of not feeling well earlier in the day, and two days earlier he had been involved in another fire where he took in a great deal of smoke.

INCIDENT 37

Camp Hill Fire Department, Camp Hill, Pennsylvania
On May 30th, Firefighter William Frank had a heart attack after returning from a heat exchanger fire at a mall.

INCIDENT 38

Globe Ranger District (USFS), Globe, Arizona
On June 9th, Firefighter Michelle Smith disappeared during a training run and was found dead twenty-six hours later. According to the autopsy report, she died of heat exhaustion and dehydration. There was no sign of struggle or foul play.

INCIDENT 39

Cameron Fire Department, Cameron, New York
On June 19th, Firefighter Rex Hoad died from injuries from a motor vehicle accident that occurred while returning from a service call.

INCIDENT 40

Poplar Springs Fire Department, Moore, South Carolina
On June 23rd, Firefighters Steele and Harmon were killed when they lost control of their truck, ran off the road, overturned, and hit a tree. The call they were responding to turned out to be a false alarm.

INCIDENT 41

Dillon County Fire Department, Dillon, South Carolina
On June 24th, Firefighter Ronald Lupo was responding to a field fire at approximately 7:10 p.m. While enroute, his vehicle was struck on the right front side by an oncoming van. He was rushed to the hospital and died later that night from internal injuries.

INCIDENT 42

Elizabeth Volunteer Fire Department, Elizabeth, West Virginia
On June 29th, Firefighter Robert Bibbee was hauling drinking water to families and homes in a rural area to raise funds for a fire department event when he suffered a heart attack.

INCIDENT 43

Cameron Volunteer Fire Department, Cameron, West Virginia
Captain Parsons, a licensed pyrotechnician, was killed at the annual 4th of July fireworks display that is sponsored by the Cameron Fire Department. A 6-inch round prematurely detonated on the ground causing a piece of metal/wood to strike Parsons in the head. His brother was also injured.

INCIDENT 44

Relief Hose Company No#2, Raritan Borough, New Jersey
On July 11th, Firefighter Bruce Lindner died from cardiac arrest during a vehicle extrication drill

INCIDENT 45

Morgan County Fire Department, Madison, Georgia
On July 11th, Firefighter George Crane Jr., died in a motor vehicle accident while responding in his personal vehicle to a 911 call.

INCIDENT 46

Holyoke Fire Department, Holyoke, Massachusetts
On July 13th, Firefighter Arthur Petit died due to cardiac arrest while searching for victims at a multiple-family dwelling fire. The firefighter's crew was ordered to search the interior of the third floor (fire floor). While searching and ventilating, Firefighter Petit collapsed on the porch of one of the apartments. He was not revived despite the attempts of fellow firefighters who performed CPR immediately.

INCIDENT 47

Pigeon Township Volunteer Fire Department, Dale, Indiana
On July 21st, Firefighter Donald Raibley was responding to a residential house fire at four in the morning when he had a seizure and drove his car over a dam into a lake, where he drowned.

INCIDENT 48

Sligo Volunteer Fire Department, Sligo, Pennsylvania
On July 27th, Firefighter Kris Sherman died from injuries resulting from an overturned pumper during a response to an incident.

INCIDENT 49

Stokes-Rockingham V.F.D., Pine Hall, North Carolina
On July 28th, Firefighter Guyer was responding in his personal vehicle to a transformer fire when his truck hydroplaned on the wet road and collided with an oncoming truck.

INCIDENT 50

Lagro Township Volunteer Fire Department, Wabash, Indiana
On August 6th, Firefighter Swan responded to a motor vehicle accident involving a car and a motorcycle. The driver of the motorcycle ran into the boat that the car was towing. The driver of the car proceeded to shoot the driver of the motorcycle, two bystanders, and the firefighter who arrived on the scene.

INCIDENT 51

Fruitland Fire Department, Fruitland, Utah
On August 8th, Firefighter Norman Ray suffered cardiac arrest due to overexertion at a grass fire.

INCIDENT 52

Addison Fire Department, Vergennes, Vermont
On August 8th, a fire broke out at the barn of Firefighter Floyd Birchmore. He called the fire department and started to lead the animals out of the barn. The first engine arrived and Firefighter Birchmore began pulling the hose off the truck. Shortly after, he collapsed and died of a heart attack.

INCIDENT 53

Surfside Volunteer Fire Department, Freeport, Texas
On August 8th, Firefighter Mac McGinnis was the only responder to a electrical pole fire that occurred mid-day. As he began deploying a hoseline, the pole broke in half, bringing the charged power lines down and electrocuting him. Pole fires occur during drought conditions that allow encrusted salt to cover insulators. There were 25 of these fires previously that summer.

INCIDENT 54

Metal Township Volunteer Fire & Ambulance Company, Fannettsburg, Pennsylvania
On August 12th, Chief Bricker suffered a fatal attack while providing patient care on an ambulance crew. Despite the efforts of his own and a neighboring ambulance crew, he was pronounced dead upon arrival at the hospital.

INCIDENT 55

Union Township Fire Department, Union Township, New Jersey
On August 24th, Deputy Chief Leslie Hendricks died as a result of cardiac arrest that was connected to an earlier fire incident at a Burger King. The chief was supervising a crew when a cloud of gas vapors engulfed him. He began having trouble breathing and was sent to the hospital. After staying two days, he returned home and died ten days later.

INCIDENT 56

Gerton Fire Department, Gerton, North Carolina
On August 21st, Firefighter Leonard Coulter died as a result of cardiac arrest while responding to a motor vehicle accident.

INCIDENT 57

Bureau of Land Management, Reno, Nevada
On August 25th, Firefighter John Gray was repairing a water tender between fires when he died of a heart attack. After he collapsed, a fellow crew member began CPR and called an ambulance.

INCIDENT 58

Harahan Fire Department, Harahan, Louisiana
On August 26th, Lieutenant Lawrence Roche had a heart attack at the scene of a structure fire.

INCIDENT 59

South Portland Fire Department, South Portland, Maine
On August 27th, Captain Robert Wallingford died from a heart attack while directing engine company operations at the scene of a four-alarm fire in a welding supply company.

INCIDENT 60

Bahama Volunteer Fire Department, Bahama, North Carolina
On September 4th, firefighters were responding to a call when a tree fell across the roadway and struck the brush truck. The accident killed Firefighter Rick Dorsey and injured one other.

INCIDENT 61

Powell Volunteer Fire Department, Powell, Wyoming
On September 10th, Assistant Chief Bruce Honstain was attempting to rescue his son from a motor vehicle accident when they were both electrocuted and died.

INCIDENT 62

Rising Sun Fire Department, Rising Sun, Maryland
On September 14th, volunteer Firefighter Sam Strall collapsed and died of a heart attack during a fund-raiser at the firehouse.

INCIDENT 63

Baltic Fire & Rescue Department, Baltic, Ohio
On September 18th, volunteer Firefighter Jeffrey Renner had arrived at his regular job, when he was informed of a fire in the paint shed. He was leaving to drive to the station to get his gear when he suffered a heart attack as he was getting to his car.

INCIDENT 64

Springdale Fire Department, Springdale, Ohio
On September 18th, Firefighter Henry Scott suffered a fatal heart attack while at a live burn training exercise.

INCIDENT 65

Birmingham Fire and Rescue Service, Birmingham, Alabama
On September 20th, Firefighter William Reid died of cardiac arrest at required annual fitness test (running & walking). He suffered an acute MI and died after being taken to the hospital.

INCIDENT 66

Herrin Fire Department, Herrin, Illinois
On September 29th, volunteer Firefighter Kevin Reveal died while fighting a fire at a commercial 2-story, vacant, boarded up, wooden building. He was opening up a boarded window that was opposite from where the fire was located when the wall collapsed, killing him and injuring several others.

INCIDENT 67

Mt. Pleasant Rural Fire, Columbia, Tennessee
On October 12th, Firefighter Clark Derryberry died in a motor vehicle accident while returning home from a barn fire. The barn fire was the last one out of a series of four.

INCIDENT 68

Jefferson Parish EBC Fire Dept., Jefferson, Louisiana
On October 13th, Firefighter Keith Boudoin was preparing to enter a structure fire for the third time to look for trapped victims when he suffered a fatal heart attack. He was immediately taken to the hospital where he was pronounced dead.

INCIDENT 69

Cowlesville Volunteer Fire Company, Cowlesville, New York
On October 15th, Assistant Chief Karl Schmidt died from an apparent massive coronary after attending an EMS training event. After the event, the chief went immediately to the hospital after arriving at home. The chief died on the way to the hospital.

INCIDENT 70

West Etowah VFD, Altoona, Alabama
On October 18th, Firefighter Martha Ann Bice was cutting firebreaks at a brush fire when she experienced chest pain and collapsed. She was taken to the hospital where they determined that she had a heart attack. While in the hospital they performed a triple bypass surgery. She went home from the hospital and died a couple of days later from complications.

INCIDENT 71

Westminster Fire Department, Westminster, Maryland
On October 19th, Firefighter Eugene Bauerline suffered a fatal heart attack at the fire department. He had been on duty all morning at the station cooking for the fire department fund-raiser. At about 11:00 a.m., he left to

direct traffic for a college homecoming. This was part of his duty for the fire police, which is a division of the fire department. He then came back to the fire department for breakfast and later went into cardiac arrest.

INCIDENT 72

Glassy Mountain Fire Department, Landrum, South Carolina
On October 24th, Captain Jackson Capps died in a motor vehicle accident when his fire truck was struck by a dump truck while responding to a grass fire. The Captain had just finished working the third shift at the electric plant when the call for a grass fire came out. After picking up a fire truck at the fire department, he pulled into a dump truck's path after driving less than 100 yards. The call turned out to be a false alarm.

INCIDENT 73

Cedar Falls Fire Department, Cedar Falls, Iowa
On October 24th, Firefighter Jack Grosse suffered cardiac arrhythmia and died while asleep in his quarters.

INCIDENT 74

Blauvelt Volunteer Fire Company, Blauvelt, New York
On October 26th, Firefighter Albert DeFlumere died of smoke inhalation at a residential structure fire when he returned inside to rescue his son.

INCIDENT 75

Portage Fire Department, Indiana
On October 26th, Firefighter Frank Gilbert died from complications due to MVA while transporting patient to hospital.

INCIDENT 76

Upper Gwynedd Township Fire Department, West Point, Pennsylvania
On November 9th, Firefighter John Bryant was fatally injured in a motor vehicle accident while responding to an alarm. The firefighter was on his way to the station when his vehicle was hit from behind at high rate of speed.

INCIDENT 77

Sharptown Volunteer Fire Department, Sharptown, Maryland
On November 9th, volunteer Firefighter Steve Trice stopped at a motor vehicle accident and was struck by a passing vehicle while attempting to extricate a victim.

INCIDENT 78

Tonawanda Fire Department, Tonawanda, New York
On November 12th, Captain Walter Schwinger Jr. died of a pulmonary embolism while asleep in the bunkroom while on duty. The crew received a call and went back in to check on the captain when he did not get up.

INCIDENT 79

Anne Arundel County Fire Department, Millersville, Maryland
On November 12th, Firefighter William Chambers collapsed and died of a heart attack during response to medical call.

INCIDENT 80

Highview Fire District, Louisville, Kentucky
On November 24th, Firefighter Donald Manuel suffered a fatal heart attack upon arrival at the scene of a church fire.

INCIDENT 81

Branford Fire Department, Branford, Connecticut
On November 27th, Firefighter Edward Ramos was killed in a warehouse fire at Floors and More, Inc. after the roof collapsed, trapping him and two other firefighters inside. Despite having his SCBA facepiece knocked off in the collapse, Firefighter Ramos stayed on the line and knocked down the fire so his comrades could escape.

INCIDENT 82

Houston Volunteer Fire Department, Houston, Texas
On December 4th, District Chief Ruben Lopez was killed in a residential structure fire while attempting to rescue one of the house's occupants. The firefighter and the victim were caught in a flashover. Both the firefighter and the victim were killed.

INCIDENT 83

Somers Fire Department, Somers, Connecticut
On December 8th, Firefighter Craig Arnone was electrocuted when his SCBA tank came into contact with a downed power line carrying 23,000 volts at a residential structure fire. A snowstorm was responsible for the

power line being down. Firefighters thought the electrical power to the area was shut off when they came to the house.

INCIDENT 84

Chicago Fire Department, Chicago, Illinois
On December 21st, Firefighter Stanley Scott suffered a fatal heart attack after hooking up a hydrant at the scene of a structure fire. CPR was initiated on the scene, but firefighter Scott was pronounced dead later at a hospital.

INCIDENT 85

Boston Fire Department, Boston, Massachusetts
On December 21st, Firefighter James A. Ellis died as a result of injuries sustained after falling approximately 20 feet down a fire pole on the way to a call. The presence of water from possibly a sink is listed as the cause of the fall. The fall caused severe head trauma and neurological damage.

INCIDENT 86

Stroh Volunteer Fire Department, Stroh, Indiana
On December 21st, Firefighter Laura Halsey was driving to the hospital with a patient from an automobile wreck, when a car struck them head on around 4:30 a.m. (no headlights and in wrong lane). All the occupants inside the car were killed.

INCIDENT 87

Burnet Volunteer Fire Department, Burnet, Texas
On December 23rd, Firefighter James Warick was struck by a vehicle while directing traffic at an incident.

INCIDENT 88

Forsyth County Fire Department, Cumming, Georgia
On December 27th, Firefighter Chesney was killed while advancing a hoseline to the upper floor of a three-story condo fire when the roof collapsed due to unseen fire spread. The two firefighters with him were able to escape.

INCIDENT 89

Kimball Township Fire Department, Hurley, Wisconsin
On December 29th, Chief Raymond Emmrich suffered a heart attack while responding to a dwelling fire. At the time of the heart attack, he was driving a pumper. He went about a block before driving into a snow bank.

Internet Resources of Interest to Safety and Health Managers

The following is by no means an exhaustive list of the volumes of safety and health-related information on the Internet. However by any measure, it is a very good start. Many if not all of the sites listed have links available that will lead the user to other similar sites.

American Red Cross Home Page
http://www.crossnet.org

American National Standards Institute
http://www.ansi.org

American Society of Safety Engineers
http://www.best.com/~assegsjc

Centers for Disease Control and Prevention
http://www.cdc.gov

Chemical Emergency Preparedness and Prevention Office
http://www.epa.gov/swercepp/

Emergency Medicine and Primary Care Home Page by EMBBS
http://www.njnet.com/~embbs/index.html

Emergency Services Registry & Search Site
http://www.district.north-van.bc.ca/admin/depart/fire/ffsearch/mainmenu.cfm

Emergency Net
http://www.emergency.com

Fire Engineering
http://www.fire-eng.com

Federal Emergency Management Agency
http://www.fema.gov

FireNet Information Network
http://online.anu.edu.au/Forestry/fire/firenet.html

Fire Protection Engineering Home Page
http://www.enfp.umd.edu/

Fire Department Safety Officers Association
http://www.fdsoa.org

Fire Marshals Association of North America
http://www.nfpa.org/pfp_div/fmana.html

Hazardous and Medical Waste Program
http://chppm-meis.apgea.army.mil

Industrial Safety & Hygiene News Magazine
http://www.SafetyOnline.net/ishn/ishn.htm

Injury Control Resource Information Network
http://www.injurycontrol.com/icrin

International Association of Fire Chiefs
http://www.iafc.org

International Association of Fire Fighters
http://www.iaff.org

IAFC Volunteer Chief Officers Section
http://www.vcos.org

National Advisory Committee for Acute Exposure Guideline Levels for Hazardous Substances
http://www.epa.gov/fedrgstr/EPA-TOX/1997/October/Day-30/t28642.htm

National Association of Emergency Vehicle Technicians
http://www.naevt.org

National Fire Protection Association
http://www.nfpa.org

National Association of Fire Equipment Distributors
http://www.nafed.org

National Institute for Computer-Assisted Reporting Data Services: OSHA
http://www.nicar.org/data/osha

National Safety Council
http://www.cais.net/nsc

National Volunteer Fire Council
http://www.nvfc.org

National Institute for Occupational Safety and Health
http://www.cdc.gov/niosh/homepage.html

National Institute for Standards and Technology
http://www.nist.gov/

Oklahoma State University—Fire Protection Programs
http://www.fireprograms.okstate.edu/index.htm

Safety Link
http://www.safetylink.com

Safety List Internet Resource Guide
http://www.halcyon.com/ttrieve/carolla.html

SafetyOnline
http://www.safetyonline.net

Safety Technology Institute
http://willow.sti.jrc.it

Safety, Occupational Health, Fire Prevention, and Emergency Services
http://www.acq.osd.mil/ens/sh

U.S. Fire Administration
http://www.usfa.fema.gov/

WPI Center for Firesafety Studies & Fire Protection Engineering
http://www.wpi.edu/Academics/Depts/Fire/fire_protection.html

WWW Emergency Sites
http://www.nw.com.au/%7Ebolin/emlist.html

Reprint of the Document Issued by the International Association of Fire Chiefs and the International Association of Fire Fighters in 1998 Regarding the Implementation of the 2-In/2-Out Rule

UNITED STATES DEPARTMENT OF LABOR OCCUPATIONAL SAFETY AND HEALTH ADMINISTRATION FIRE FIGHTERS' TWO-IN/TWO-OUT REGULATION

The federal Occupational Safety and Health Administration (OSHA) recently issued a revised standard regarding respiratory protection. Among other changes, the regulation now requires that interior structural fire fighting procedures provide for at least two fire fighters inside the structure. Two fire fighters inside the structure must have direct visual or voice contact between each other and direct, voice or radio contact with fire fighters outside the structure. This section has been dubbed the fire fighters' "two-in/two-out" regulation. The International Association of Fire Fighters and the International Association of Fire Chiefs are providing the following questions and answers to assist you in understanding the section of the regulation related to interior structural fire fighting.

1. *What is the federal OSHA Respiratory Protection Standard?*

In 1971, federal OSHA adopted a respiratory protection standard requiring employers to establish and maintain a respiratory protection program for their respiratory-wearing employees. The revised standard strengthens some requirements and eliminates duplicative requirements in other OSHA health standards.

The standard specifically addresses the use of respirators in immediately dangerous to life or health (IDLH) atmospheres, including interior structural fire fighting. OSHA defines structures that are involved in fire beyond the incipient stage as IDLH atmospheres. In these atmospheres, OSHA requires that personnel use self-contained breathing apparatus (SCBA), that a minimum of two fire fighters work as a team inside the structure, and that a minimum of two fire fighters be on standby outside the structure to provide assistance or perform rescue.

2. *Why is this standard important to fire fighters?*

This standard, with its two-in/two-out provision, may be one of the most important safety advances for fire fighters in this decade. Too many fire fighters have died because of insufficient accountability and poor communications. The standard addresses both and leaves no doubt that two-in/two-out requirements must be followed for fire fighter safety and compliance with the law.

3. *Which fire fighters are covered by the regulations?*

The federal OSHA standard applies to all private sector workers engaged in fire fighting activities through industrial fire brigades, private incorporated fire companies (including the "employees" of incorporated volunteer companies and private fire departments contracting to public jurisdictions) and federal fire fighters. In 23 states and 2 territories, the state, not the federal government, has responsibility for enforcing worker health and safety regulations. These "state plan" states have earned the approval of federal OSHA to implement their own enforcement programs. These states must establish and maintain occupational safety and health programs for all public employees that are as effective as the programs for private sector employees. In addition, state safety and health regulations must be at least as stringent as federal OSHA regulations. Federal OSHA has no direct enforcement authority over state and local governments in states that do not have state OSHA plans.

All professional career fire fighters, whether state, county, or municipal, in any of the states or territories where an OSHA plan agreement is in effect, have the protection of all federal OSHA health and safety standards, **including the new respirator standard and its requirements for fire fighting operations.** The following states have OSHA-approved plans and must enforce the two-in/two-out provision for all fire departments.

Alaska	Kentucky	North Carolina	Virginia
Arizona	Maryland	Oregon	Virgin Islands
California	Michigan	Puerto Rico	Washington
Connecticut	Minnesota	South Carolina	Wyoming
Hawaii	Nevada	Tennessee	
Indiana	New Mexico	Utah	
Iowa	New York	Vermont	

A number of other states have adopted, by reference, federal OSHA regulations for public employee fire fighters. These states include Florida, Illinois and Oklahoma. In these states, the regulations carry the force of state law.

Additionally, a number of states have adopted NFPA standards, including NFPA 1500, *Standard for Fire Department Occupational Safety and Health Program*. The 1997 edition of NFPA 1500 now includes requirements corresponding to OSHA's respiratory protection regulation. Since the NFPA is a private consensus standards organization, its recommendations are preempted by OSHA regulations that are more stringent. In other words, the OSHA regulations are the minimum requirement where they are legally applicable. There is nothing in federal regulations that "deem compliance" with any consensus standards, including NFPA standards, if the consensus standards are less stringent.

It is unfortunate that all U.S. and Canadian fire fighters are not covered by the OSHA respiratory protection standard. However, we must consider the two-in/two-out requirements to be the minimum acceptable standard for safe fire ground operations for all fire fighters when self-contained breathing apparatus is used.

4. *When are two-in/two-out procedures required for fire fighters?*

OSHA states that "once fire fighters begin the interior attack on an interior structural fire, the atmosphere is assumed to be IDLH and paragraph **29 CFR 1910.134(g)(4)** [two-in/two-out] applies." OSHA defines interior structural fire fighting "as the physical activity of fire suppression, rescue or both inside of buildings or enclosed structures which are involved in a fire situation beyond the incipient stage." OSHA further defines an incipient stage fire in **29 CFR 1910.155(c)(26)** as a "fire which is in the initial or beginning stage and which can be controlled or extinguished by portable fire extinguishers, Class II standpipe or small hose systems without the need for protective clothing or breathing apparatus." Any structural fire beyond incipient stage is considered to be an IDLH atmosphere by OSHA.

5. *What respiratory protection is required for interior structural fire fighting?*

OSHA requires that all fire fighters engaged in interior structural fire fighting must wear SCBAs. SCBAs must be NIOSH-certified, positive pressure, with a minimum duration of 30 minutes. **[29 CFR 1910.156(f)(1)(ii)]** and **[29 CFR 1910.134(g)(4)(iii)]**

6. *Are all fire fighters performing interior structural fire fighting operations required to operate in a buddy system with two or more personnel?*

Yes. OSHA clearly requires that all workers engaged in interior structural fire fighting operations beyond the incipient stage use SCBA and work in teams of two or more. **[29 CFR 1910.134(g)(4)(i)]**

7. *Are fire fighters in the interior of the structure required to be in direct contact with one another?*

Yes. Fire fighters operating in the interior of the structure must operate in a buddy system and maintain voice or visual contact with one another at all times. This assists in assuring accountability within the team. **[29 CFR 1910.134(g)(4)(i)]**

8. *Can radios or other means of electronic contact by substituted for visual or voice contact, allowing fire fighters in an interior structural fire to separate from their "buddy" or "buddies"?*

No. Due to the potential of mechanical failure or reception failure of electronic communication devices, radio contact is not acceptable to replace visual or voice contact between the members of the "buddy system" team. Also, the individual needing rescue may not be physically able to operate an electronic device to alert other members of the interior team that assistance is needed.

Radios can and should be used for communications on the fire ground, including communications between the interior fire fighter team(s) and exterior fire fighters. They cannot, however, be the sole tool for accounting for one's partner in the interior of a structural fire. **[29 CFR 1910.134(g)(4)(i)]** **[29 CFR 1910.134(g)(3)(ii)]**

9. *Are fire fighters required to be present outside the structural fire prior to a team entering and during the team's work in the hazard area?*

Yes. OSHA requires at least one team of two or more properly equipped and trained fire fighters be present outside the structure before any team(s) of fire fighters enter the structural fire. This requirement is intended to assure that the team outside the structure has the training, clothing and equipment to protect themselves and, if necessary, safely and effectively rescue fire fighters inside the structure. For high-rise operations, the team(s) would be staged below the IDLH atmosphere. **[29 CFR 1910.134(g)(3)(iii)]**

10. *Do these regulations mean that, at a minimum, four individuals are required, that is, two individuals working as a team in the interior of the structural fire and two individuals outside the structure for assistance or rescue?*

Yes. OSHA requires that a minimum of two individuals, operating as a team in direct voice or visual contact, conduct interior fire fighting operations utilizing SCBA. In addition, a minimum of two individuals who are properly equipped and trained must be positioned outside the IDLH atmosphere, account for the interior team(s) and remain capable of rapid rescue of the interior team. The outside personnel must at all times account for and be available to assist or rescue members of the interior team. **[29 CFR 1910.134(g)(4)]**

11. *Does OSHA permit the two individuals outside the hazard area to be engaged in other activities, such as incident command or fire apparatus operation (for example, pump or aerial operators)?*

OSHA requires that one of the two outside person's function is to account for and, if necessary, initiate a fire fighter rescue. Aside from this individual dedicated to tracking interior personnel, the other designated person(s) is permitted to take on other roles, such as incident commander in charge of the emergency incident, safety officer or equipment operator. However, the other designated outside person(s) cannot be assigned tasks that are critical to the safety and health of any other employee working at the incident.

Any task that the outside fire fighter(s) performs while in standby rescue status must not interfere with the responsibility to account for those individuals in the hazard area. Any task, evolution, duty, or function being performed by the standby individual(s) must be such that the work can be abandoned, without placing any employee at additional risk, if rescue or other assistance is needed. **[29 CFR 1910.134(g)(4)(Note 1)]**

12. *If a rescue operation is necessary, must the buddy system be maintained while entering the interior structural fire?*

Yes. Any entry into an interior structural fire beyond the incipient stage, regardless of the reason, must be made in teams of two or more individuals. **[29 CFR 1910.134(g)(4)(i)]**

13. *Do the regulations require two individuals outside for* **each** *team of individuals operating in the interior of a structural fire?*

The regulations do not require a separate "two-out" team for each team operating in the structure. However, if the incident escalates, if accountability cannot be properly maintained from a single exposure, or if rapid rescue becomes infeasible, additional outside crews must be added. For example, if the involved structure is large enough to require entry at different locations or levels, additional "two-out" teams would be required. **[29 CFR 1910.134(g)(4)]**

14. *If four fire fighters are on the scene of an interior structural fire, is it permissible to enter the structure with a team of two?*

OSHA's respiratory protection standard is not about counting heads. Rather, it dictates functions of fire fighters prior to an interior attack. The entry team must consist of at least two individuals. Of the two fire fighters outside, one must perform accountability functions and be immediately available for fire fighter rescue. As explained above, the other may perform other tasks, as long as those tasks do not interfere with the accountability functions and can be abandoned to perform fire fighter rescue. Depending on the operating procedures of the fire department, more than four individuals may be required. **[29 CFR 1910.134(g)(4)(i)]**

15. *Does OSHA recognize any exceptions to this regulation?*

OSHA regulations recognize deviations to regulations in an emergency operation where immediate action is necessary to save a life. For fire department employers, initial attack operations must be organized to ensure that adequate personnel are at the emergency scene prior to any interior at a structural fire. If initial attack personnel find a **known** life-hazard situation where immediate action could prevent the loss of life, deviation from the two-in/two-out standard may be permitted, as an exception to the fire department's organizational plan.

However, such deviations from the regulations must be **exceptions** and not defacto standard practices. In fact, OSHA may still issue "de minimis" citations for such deviations from the standard, meaning that the citation will not require monetary penalties or corrective action. The exception is for a known life rescue only, not for standard search and rescue activities. When the exception becomes the practice, OSHA citations are authorized. **[29 CFR 1910.134(g)(4)(Note 2)]**

16. *Does OSHA require employer notification prior to any rescue by the outside personnel?*

Yes. OSHA requires the fire department or fire department designee (i.e., incident commander) be notified prior to any rescue of fire fighters operating in an IDLH atmosphere. The fire department would have to provide any additional assistance appropriate to the emergency, including the notification of on-scene personnel and incoming units. Additionally, any such actions taken in accordance with the "exception" provision should be thoroughly investigated by the fire department with a written report submitted to the Fire Chief. **[29 CFR 1910.134(g)(3)(iv)]**

17. *How do the regulations affect fire fighters entering a hazardous environment that is not an interior structural fire?*

Fire fighters must adhere to the two-in/two-out regulations for other emergency response operations in any IDLH, potential IDLH, or unknown atmosphere. OSHA permits one standby person **only** in those IDLH environments in fixed workplaces, not fire emergency situations. Such sites, in normal operating conditions, contain only hazards that are known, well characterized, and well controlled. **[29 CFR 1910.120(q)(3)(vi)]**

18. *When is the new regulation effective?*

The revised OSHA respiratory protection standard was released by the Department of Labor and published in the Federal Register on January 8, 1998. It is effective on April 8, 1998.

"State Plan" states have six months from the release date to implement and enforce the new regulations.

Until the April 8 effective date, earlier requirements for two-in/two-out are in effect. The formal interpretation and compliance memo issued by James W. Stanley, Deputy Assistant Secretary of Labor, on May 1, 1995 and the compliance memo issued by Assistant Secretary of Labor Joe Dear on July 30, 1996 establish that OSHA interprets the earlier 1971 regulation as requiring two-in/two-out. **[29 CFR 1910.134(n)(1)]**

19. *How does a fire department demonstrate compliance with the regulations?*

Fire departments must develop and implement standard operating procedures addressing fire ground operations and the two-in/two-out procedures to demonstrate compliance. Fire department training programs must ensure that fire fighters understand and implement appropriate two-in/two-out procedures. **[29 CFR 1910.134(c)]**

20. *What can be done if the fire department does not comply?*

Federal OSHA and approved state plan states must ". . . assure so far as possible every working man and woman in the Nation safe and healthful working conditions." To ensure such protection, federal OSHA and states with approved state plans are authorized to enforce safety and health standards. These agencies must investigate complaints and conduct inspections to make sure that specific standards are met and that the workplace is generally free from recognized hazards likely to cause death or serious physical harm.

Federal OSHA and state occupational safety and health agencies must investigate written complaints signed by current employees or their representatives regarding hazards that threaten serious physical harm to workers. By law, federal and state OSHA agencies do not reveal the name of the person filing the complaint, if he or she so requests. Complaints regarding imminent danger are investigated even if they are unsigned or anonymous. For all other complaints (from other than a current employee, or unsigned, or anonymous), the agency may send a letter to the employer describing the complaint and requesting a response. It is important that an OSHA (either federal or state) complaint be in writing.

When an OSHA inspector arrives, he or she displays official credentials and asks to see the employer. The inspector explains the nature of the visit, the scope of the inspection and applicable standards. A copy of any employee complaint (edited, if requested, to conceal the employee's identity) is available to the employer. An employer representative may accompany the inspector during the inspection. An authorized representative of the employees, if any, also has the right to participate in the inspection. The inspector may review records, collect information and view work sites. The inspector may also interview employees in private for additional information. Federal law prohibits discrimination in any form by employers against workers because of anything that workers say or show the inspector during the inspection or for any other OSHA protected safety-related activity.

Investigations of imminent danger situations have top priority. An imminent danger is a hazard that could cause death or serious physical harm immediately, or before the danger can be eliminated through normal enforcement procedures. Because of the hazardous and unpredictable nature of the fire ground, a fire department's failure to comply with the two-in/two-out requirements creates an imminent danger and the agency receiving a related complaint must provide an immediate response. If inspectors find imminent danger conditions, they will ask for immediate voluntary correction of the hazard by the employer or removal of endangered employees from the area. If an employer fails to do so, federal OSHA can go to federal district court to force the employer to comply. State occupational safety and health agencies rely on state courts for similar authority.

Federal and state OSHA agencies are required by law to issue citations for violations for safety and health standards. The agencies are not permitted to issue warnings. Citations include a description of the violation, the proposed penalty (if any), and the date by which the hazard must be corrected. Citations must be posted in the workplace to inform employees about the violation and the corrective action. **[29 CFR 1903.3(a)]**

It is important for labor and management to know that this regulation can also be used as evidence of industry standards and feasibility in arbitration and grievance hearings on fire fighter safety, as well as in other civil or criminal legal proceedings involving injury or death where the cause can be attributed to employer failure to implement two-in/two-out procedures. Regardless of OSHA's enforcement authority, this federal regulation links fire ground operations with fire fighter safety.

21. *What can be done if a fire fighter does not comply with fire department operating procedures for two-in/two-out?*

Fire departments must amend any existing policies and operational procedures to address the two-in/two-out regulations and develop clear protocols and reporting procedures for deviations from these fire department policies and procedures. Any individual violating this safety regulation should face appropriate departmental action.

22. *How can I obtain additional information regarding the OSHA respirator standard and the two-in/two-out provision?*

Affiliates of the International Association of Fire Fighters may contact:

International Association of Fire Fighters
Department of Occupational Health and Safety
1750 New York Avenue, NW
Washington, DC 20006
202-737-8484
202-737-8418 (FAX)

Members of the International Association of Fire Chiefs may contact:

International Association of Fire Chiefs
4025 Fair Ridge Drive
Fairfax, VA 22033-2868
703-273-0911
703-273-9363 (FAX)

Does the new 2 in/2 out regulation apply to volunteer fire departments and volunteer fire fighters?

Yes, it is the opinion of the IAFC Volunteer Chief Officers Section (VCOS) that this regulation is applicable in all situations where interior fire fighting operations are being conducted. VCOS believes it is applicable to the vast majority of volunteer fire departments in the United States.

The VCOS supported the concept of 2 in/2 out when it was originally proposed and VCOS still supports these new regulations since they deal directly with the safety and well being of our fire fighters.

While OSHA has stated that volunteer fire departments will not be affected by the regulations, VCOS believes that this will not be the case since 25 states align with federal OSHA. The standard also applies to private incorporated fire companies including the "employees" of incorporated volunteer companies. In addition, other non-OSHA states are giving the regulations consideration under the Environmental Protection Agency, the federal entity that typically takes the lead in non-OSHA states.

The application of the 2 in/2 out rule can be argued by volunteer fire departments in many states. The reality is that 2 in/2 out is the right thing to do for the safety of our fire fighters and each volunteer fire department should seek to implement new standard operating guidelines for 2 in/2 out.

For further information on the application of 2 in/2 out to volunteers, contact:

Volunteer Chief Officers Section
International Association of Fire Chiefs
4025 Fair Ridge Drive
Fairfax, VA 22033-2868
703-273-0911
703-273-9363 (FAX)

Acronyms

AFEM	Alliance for Fire and Emergency Management
ANSI	American National Standard Institute
ASTM	American Society of Testing and Materials
BLEVE	Boiling liquid expanding vapor explosion
BSI	Body substance isolation
CDL	Commercial driver's license
CFR	Code of Federal Regulations
CISM	Critical incident stress management
CVD	Cardiovascular disease
DRDS	Division of Respiratory Disease Studies
DSHEFS	Division of Surveillance, Hazard Evaluations, and Field Studies
DSR	Division of Safety Research
EMS	Emergency medical services
EPA	Environmental Protection Agency
EVOC	Emergency vehicle operator courses
FACE	Fatality Assessment and Control Evaluation
GPS	Global positioning system
HAZWOPER	Hazardous Waste Operations and Emergency Response
HBV	Hepatitis B virus
HIV	Human immunodeficiency virus
IAFC	International Association of Fire Chiefs
IAFF	International Association of Fire Fighters
IC	Incident commander
IDLH	Immediately dangerous to life and health
IMS	Incident management system
IV	Intravenous
MVA	Motor vehicle accident
NAEVT	National Association of Emergency Vehicle Technicians
NBC	Nuclear, biological, or chemical
NFIRS	National Fire Incident Reporting System
NFPA	National Fire Protection Association
NIOSH	National Institute for Occupational Safety and Health
NIST	National Institute for Standards and Technology
NVFC	National Volunteer Fire Council
OSHA	Occupational Safety and Health Administration
PAR	Personal accountability reports
PASS	Personal alert safety system
PIA	Postincident analysis
PM	Preventive maintenance
PPE	Personal protective equipment
PSOB	Public Safety Officer's Benefit Program
RAID	Recognize, approach, identify, decide
RIC	Rapid intervention company
RPD	Recognition-primed decision
SCBA	Self-contained breathing apparatus
SLFD	St. Louis Fire Department
SOP	Standard operating procedure
USFA	United States Fire Administration
VCOS	Volunteer Chief Officer's Section

Glossary

Aerobic fitness A measurement of the body's ability to perform and utilize oxygen.

Alternative duty programs Sometimes called light duty or modified duty, these programs allow an injured employee to return to work, with restrictions for some period of time while recuperating.

Back drafts Occur as a result of burning in an oxygen-starved atmosphere. When air is introduced, the superheated gases ignite with enough force to be considered an explosion.

BLEVES Boiling liquid expanding vapor explosions; occur when heat is applied to a liquified gas container and the gas expands at a rapid rate while the container is weakened by the heat. When the container fails a BLEVE is said to occur.

Bloodborne pathogen Disease carried in blood or blood products.

Body composition A measure to show the percentage of fat in the body; there are certain published parameters for what is considered average or normal.

Cardiovascular fitness Fitness levels associated with the cardiovascular system, including the heart and circulatory system.

Civil disturbances Uprisings of civilians that often lead to hostile acts against law enforcement and emergency responders.

Code of Federal Regulations (CFR) The document that contains all of the federally promulgated regulations for all federal agencies.

Cognitive Skills learned through a mental learning process as opposed to practical learning.

Consensus standards Standards developed by consensus of industry or subject area experts, which are then published and may or may not be adopted locally. Even if not adopted as law, these standards can often be used as evidence for standard of care.

Critical incident stress Stress associated with critical incidents, such as the injury or death of a coworker or a child.

Critical incident stress management (CISM) A process for managing the short- and long-term effects of critical incident stress reactions.

Demobilization The process of returning personnel, equipment, and apparatus after an emergency has been terminated.

Fire brigades The use of trained personnel within a business or at an industry site for fire fighting and emergency response.

First in–last out The common approach often used at an emergency scene. Basically, the first arriving crews are generally the last to leave the scene.

Flashover A sudden, full involvement in flame of materials and gases within a room.

Freelancing Occurs when responders do not follow the incident plan at a scene and do what they want on their own. A failure to stay with assigned group.

Frequency How often a risk occurs or is expected to occur.

Goals Broad statements of what needs to be accomplished.

Haddon matrix A 4 × 3 matrix used to help analyze injuries and accidents in an attempt to determine processes designed to reduce them.

Heat transfer The transfer of heat through conduction, convection, radiation, and direct flame contact.

Hits Number of documents found when searching for a key word or phrase using a search engine.

Immediately dangerous to life and health (IDLH) Used by several OSHA regulations to describe a process or an event that could produce loss of life or serious injury if a responder is exposed or operates in the environment.

Incident management system (IMS) An expandable management system for dealing with a myriad of incidents to provide the highest level of accountability and effectiveness. Limits span of control and provides a framework of breaking the big job down into manageable tasks.

International Association of Fire Chiefs (IAFC) Organization of fire chiefs from the United States and Canada.

International Association of Fire Fighters (IAFF) Labor organization that represents the majority of organized firefighters in the United States and Canada.

Links Used on Internet pages so that each page may be directly tied to a page on another Internet site.

Matrix A chart used to categorize actions or events for analysis.

Mushrooming Heat and gases accumulate at the ceiling or top floor of a multistory building then back down. Can be prevented with vertical ventilation.

National Association of Emergency Vehicle Technicians (NAEVT) An organization that bestows professional certification in many areas for persons involved in emergency service vehicle maintenance.

National Fire Incident Reporting System (NFIRS) The uniform fire incident reporting system for the United States; the data from this report is analyzed by the United States Fire Administration.

NFPA 1403 The National Fire Protection Association's consensus standard, *Live Fire Training Evolutions.*

NFPA 1500 The National Fire Protection Association's consensus standard *Fire Department Occupational Safety and Health Program.*

Occupational Safety and Health Administration (OSHA) Federal agency tasked with the responsibility for the occupational safety of employees.

OSHA 200 A list of all on-the-job injuries within a given work site. Required to be posted for all to see.

Outcome Evaluation An evaluation that answers the question, "Did the program meet the expected goals?"

Peer defusing The concept of using a trained person from the same discipline to talk to a emergency responder after a critical incident occurs as a means to allow the responder to talk about his or her feelings about the event in a nonthreatening environment.

Personal alert safety system (PASS) A device that produces a high-pitched audible alarm when the wearer becomes motionless for some period of time; useful to attract rescuers to a downed firefighter.

Personnel accountability reports (PAR) A verbal or visual report to incident command or to the accountability officer regarding the status of operating crews. Should occur at specific time intervals or after certain tasks have been completed.

Postincident analysis (PIA) A critical review of the incident after it occurs. The postincident analysis should focus on improving operational effectiveness and safety.

Preventive maintenance (PM) program An ongoing program for maintenance on vehicles. Designed to provide routine care, oil changes and the like, as well as catch minor problems before they become major ones.

Process Evaluation The evaluation of the various processes associated with a program or task that is on going.

Rapid interview companies (RIC) An assignment of a group of rescuers with the sole purpose of rapid deployment to reports of operating personnel in trouble or missing.

Regulations Rules or laws promulgated at the federal, state, or local level with a requirement to comply.

Rehabilitation The group of activities that ensures responders' health and safety at an incident scene. May include rest, medical surveillance, hydration, and nourishment.

Risk control A common approach to risk management where measures and processes are implemented to help control the number and the severity of losses or consequences of risk to the organization.

Risk management Processes and programs put in place to minimize risks and reduce the consequences when an accident occurs.

Risk transfer The process of letting someone else assume the risk; for example, buying auto insurance transfers the consequences of an accident to the insurance company.

Risks The resultant outcome of exposure to a hazard.

Rollover The rolling of flame under the ceiling as a fire progresses to the flashover stage.

Search Engines Programs on the Internet that allow a user to search the entire Internet for key words or phrases.

Severity How severe the result is when a risk occurs.

Standard of care The concept of what a reasonable person with similar training and equipment would do in a similar situation.

Standard operating procedures (SOP) Sometimes called standard operating guidelines, these are department-specific operational procedures, policies, and rules made to assist with standardized actions at various situations.

Standardized Apparatus Apparatus that has exactly the same operation and layout of other similar apparatus in a department. For example, all of the department's pumpers would be laid out the same, operate the same, and have the same equipment. Useful for the situations when crews must use another crew's apparatus.

Standards Often developed through the consensus process, standards are not mandatory unless adopted by a governmental authority.

Stress The body's reaction to an event. Not all stress is bad, in fact some level of stress is necessary to get a person to perform, for example, the stress associated with a report that is due is often the motivating factor in doing it.

United States Fire Administration Agency under the Federal Emergency Management Agency that directs and produces fire programs, research, and education.

Index

A

B

C

D

E

F

G

H

I

L

M

N

O

P

R

S